The Musician's Brain

Does It Recover from Trauma Better Than Others?

Carol Shively Mizes

PUBLISHER'S PREFACE

BookBaby

7905 North Route 130

Pennsauken, NJ 08110

www.bookbaby.com

877.961.6878

The Musician's Brain

Does it Recover From Trauma Better Than Others?

First paperback printing, 2017

Print ISBN: 978-1-48359-385-2

eBook ISBN: 978-1-48359-386-9

Musicians in Trauma

When a musician suffers major brain trauma,

How differently does he or she heal?

Did active instrumental playing or lifelong singing

Scaffold their brain as if in steel?

If a musician suffers a trauma and receives Music Therapy

During their road to wellness,

Can we spark neuro-connections faster or stronger

Than with one who wasn't so musically zealous?

Carol Shively Mizes, MT-BC
Neurologic Music Therapy Fellow

FOREWORD

Why I Decided to Write This Book

I am writing this book to share the wonderful recovery stories of musicians who have suffered severe trauma and with whom I've had the honor to work.

These individuals were collegiate, amateur or professional musicians. Their injuries were so severe, some were not expected to live much less recover well enough to return to their normal lifestyles.

You will read about people who suffered ruptured brain aneurysms, traumatic brain injuries from car or bike accidents as well as from falls and you will appreciate that when a person suffers a trauma, their family and friends suffer as well.

I am trusting that these stories will be an inspiration to families who are experiencing similar trauma, to know there is hope. I want others to realize the extraordinary care that is provided by Cleveland's MetroHealth System. I hope to help people understand more fully the impact that music therapy plays in the role of the recovery process. Properly applied *Neurologic Music Therapy* (NMT) techniques can be very powerful for those with brain and spinal trauma as well as stroke. I am, however, particularly amazed with how well musicians recover.

After 35 years in music therapy including the last twelve years at the MetroHealth Rehabilitation Institute of Ohio, I have been a part of the rehabilitation team of therapists and have seen numerous musicians recover remarkably well. I would like to acknowledge all the other therapists, physicians and nurses who played a significant role in the recovery process of our patients.

We now know that a musician's brain is very different than a non-musician. Musicians have stronger and thicker neuro-fibers in different areas of the brain from playing their instruments or from singing for many years. Simply put, their brains are wired differently. I believe they have more to fall back on when suffering a brain trauma or stroke. This might be the reason I have seen musicians recover more fully or efficiently, *especially* when receiving music therapy in their rehabilitation. Preliminary research supports my position on this. Some researchers are finding that very active/ practicing musicians, such as professional orchestra musicians do not contract the degenerative brain diseases others suffer such as the different forms of dementia. *Active Music playing seems to be a protectant for the brain.*

Having said that, there needs to be much more research into music therapy and the musician's recovery process. I've provided a review of the literature near the back of this book. It is not intended to be all inclusive but an informational review, should anyone desire to delve more deeply.

After twelve years in the Rehabilitation Department, I now report to a new MetroHealth Department aptly named; *Arts in Medicine.* I am the Coordinator of Arts Therapies. This department was launched in December of 2014 with Linda Jackson, Director. Jackson has supported me in my endeavor to write this book.

My new role has allowed me to expand music therapy services to the Intensive Care Units (ICUs) where music can and does positively affect the injured brain - *even if the patient is not yet responsive.* Most of my rehabilitation patients started their recovery process in the Intensive Care Unit. Their stories along with others have inspired me to expand music therapy services into our ICUs.

Although music therapy is beneficial for anyone who has suffered a trauma or injury, this book has a special purpose; to focus on how *musicians* recover. My hope for this book is that it provides a basis for future research in this area.

Numerous hospitals around the world now include *The Arts in Medicine*. There is a reason for that. The Arts, in many ways, help patients in and out of hospitals recover better and faster. We need to come to a better understanding of the power of The Arts but this book will focus on musicians and music therapy for brain trauma recovery.

Carol Shively Mizes, MT-BC
Neurologic Music Therapy Fellow
Coordinator of Arts Therapies,
Department of Arts in Medicine
The MetroHealth System.

TABLE OF CONTENTS

DEDICATIONS

I am dedicating this book to all of my former patients and their families who have allowed me to share their wonderful stories of perseverance, hard work and a willingness to participate in music therapy. They inspired me.

I believe God put this passion in my heart as a young high school sophomore in 1974, even at a time when we didn't yet know the science of music and its true biological role in the recovery process of the injured brain and body. I have worked with many more musicians in rehabilitation who experienced spinal cord injuries or strokes, but I will not be writing about their stories at this time, even though they recovered well. I chose the people for this book based on the fact that their injuries were somewhat more *life threatening* and the fact that I have kept in touch with them over the years. This is NOT to say that the other situations were not difficult or serious.

Because the effects of music and music techniques have shown to re-wire neuro-connections, increase neuro-chemicals that stir emotions, elevate mood and decrease pain, as well as regulate biological rhythms across multi-cultural and socio-economic groups, we can actually say that

"Music Is The Universal Language"

God bless all of you!

A Special Dedication to Mrs. Dee Charlton

Thank you so much for the countless hours you spent editing this work. Your ideas and recommendations were extremely valuable. Your support in my endeavor to write and publish these stories helped me to stay the course and forge ahead.

No words can express how thankful I really am for the time and attention to detail you have provided.

Thank you Dee!

CHAPTER I

Greg, The Lake County Musician Who Collapsed Mid Performance

"By the time I got there (to the local emergency department) he was intu-bated, in grave condition and Life Flight was on their way to take him to MetroHealth Hospital. I knew there was a very good chance that he would not make it through the night. It's surreal when at 4:30 p.m. you're having a snack together and then at 7:30 p.m. he's fighting for his life."

Greg's wife, Sue, recalled her night of doom.

Greg was singing and playing guitar for a small group of people in Madison, Ohio. Suddenly, as he stood in front of the group, he collapsed.

At first, people thought it was part of his act because he just happened to be singing the song *What Will We Do with a Drunken Sailor?* They thought he was acting out for comic relief, but when he didn't respond, someone in the audience called 911. When the Emergency Medical Services arrived, they took him to the nearest emergency department, where it was quickly determined he needed to be at MetroHealth Medical Center in Cleveland, Ohio (the only verified Level One Trauma Center in the city, established in 1992). They called Metro Life Flight. It was raining when Life Flight left MetroHealth but the rain soon turned into a horrific storm and the pilot had to make the call; it was unsafe to continue flying, so they decided to call for an ambulance. Madison is approximately one hour away from Cleveland by car or ambulance. Every *second* counts with a brain trauma and now he was literally in for the ride of his life!

Sue rode with him on that long, torturous drive and didn't think he'd make it through the night. Greg was brought to the emergency room at MetroHealth where it was confirmed that he suffered a ruptured brain aneurysm causing massive bleeding. It became critical to perform immediate brain surgery to stop the bleeding, repair the ruptured aneurysm and then wait, monitoring every breath, every heartbeat and every vital sign. Waiting is the hardest, especially for the families. Greg remained in the Surgical Intensive Care Unit at MetroHealth for 23 three days.

During his time in critical care his friend, Lori Rizzo, a singer who often performed with Greg, came to visit him. He was still not responding well; he was not able to speak or do anything for himself but was able to squeeze your hand to let you know he heard you. Lori asked Greg's nurse if she could sing to him. She said *"sure"*. Lori started to sing *Desperado,* written by Glenn Frey and Don Henley from the Eagles, one of their favorite songs to perform together. Shockingly, Greg started to sing with her, in full voice and in perfect pitch! He was so loud his nurse could hear him from the next room and ran into Greg's room to find out who was singing with Lori. No one could believe it was him singing all the words and in beautiful harmony! This was the spark for recovery. After a total of 23 days, on September 6, 2011, Greg was moved to the Rehabilitation hospital.

I was walking down the hall on the brain injury unit in the Rehabilitation Department and was approached by one of our Occupational Therapists, Laura Poling. She thought this particular patient would be a good candidate for music therapy.

"Carol, can you meet with the patient in room 804, bed one? He's a musician."

When I went to his room to introduce myself, I saw a somewhat fragile man wearing a hospital gown, resting on his bed. His head was shaved due to surgery and the staples were exposed on his large incision. He was smiling and friendly but confused and quite chatty. I asked him his name and he replied, *"Greg."*

I noticed a ukulele in his room so I picked it up, gave it to him and in order to test his memory and verbal skills, I asked, *"what is this?"* He replied, *"ukulele, I played it so intimately for 35 years."* Since it was in his room, I believed him so I asked him to play something for me. He tried but whatever he was playing was NOT a song or any type of organized music I could recognize. I think even he, with a severe brain injury, knew it didn't sound quite right, so he decided to Google 'ukulele chords' on his laptop. This showed good problem solving skills. I was impressed!

He described the different types of guitars he had at home and how he used to play drums in high school and for seven years played in a band of his own. He continued (in his chatty mode) to explain how the drummer was the first guy at the gig to set up in rooms that were sometimes too small for the drum set. People complained it was too loud but you can't play drums that softly. Continuing, he informed me that as the drummer he noticed the guy standing in front, usually the singer/guitar player got all the girls. So from then on, he decided to play guitar and sing. Greg seemed happy to talk and talk about his musical experiences.

I explained that I wanted to begin seeing him the next day in music therapy and that with a regular schedule, he would gradually get stronger, feel better and re-gain his music skills as a bonus. Greg agreed to participate even though he probably didn't remember what I said a few minutes later. He just enjoyed music so much.

The next day I read his medical chart and reason for his hospitalization as I always do prior to treating patients. His name was Greg Markell, a fifty-six-year-old man diagnosed with a brain aneurysm that ruptured. He needed critical surgery to repair it. The hemorrhage caused stroke-like symptoms; memory deficits, confusion, poor judgment, physical weakness and poor balance, along with the fact that he was unaware of the severity of all these deficits. Much later after I met Greg, I heard the story of Lori Rizzo singing with him in ICU. Through all the years I've worked with those who suffer from brain injuries and strokes, I truly believe that singing a familiar and favorite song to Greg, flipped the switch on in his brain. The neurons and pathways that were accustomed to being used in this way prior to being severely injured, were awakened again.

Soon Greg was strong enough to be admitted to MetroHealth's rehab center and tolerate at least three hours of therapy per day. *"Music therapy became the generator that kept my brain fired up."* This is how Greg likes to describe his experience.

A couple of days after I met Greg, I had the opportunity to meet his wife Sue. She related his rich history in music and explained that Greg does *not* play ukulele! One of his friends brought it to the hospital for Greg to play around with. In his confused state of mind, he did not know that he didn't play ukulele, and he didn't know how the fingerings were different than on a guitar since the 'uke' only has four strings, he was used to six on the guitar. I said to Sue, *"no wonder it didn't sound so well when he played the ukulele for me."* I explained to her that often, people with brain injuries fabricate stories though they truly believe what they are saying is truthful and correct. Greg fabricated a lot in the beginning.

During our music therapy evaluation, I asked Greg many questions for different reasons;

1. I needed to see what he could remember and gauge his level of orientation.

2. I needed to have some biographical information about him to know how to plan our sessions together. It went something like this:

Carol: *"Tell me your full name please."*

Greg: *"Gregory P. Markell."*

Carol: *"What kind of music do you like?"*

Greg: Ever the funny guy, *"I like music in the key of G."*

I gave examples of genres and he told me which ones he liked; Country, Rock 'n' Roll, Polkas, Pop but mostly singer/songwriter songs. *"I'm incorporating some old Beatles' songs into my playlist. They've been received well by people sitting on stools in bars where I play."*

Carol: *"Did you ever sing anywhere else?"*

Greg; *"I sang in the Methodist youth choir where the director at church was also the Music Director at Mentor High during my high school years."*

Carol: *"Do you have pain?"*

Greg: *"Yes, there is this achy thing going on right here* (as he pointed to his surgical area,) *which I suspect was the aneurysm."*

Carol: *"Can you rate your pain from one to 10 with one being no pain and 10 the worst pain you've ever felt?"*

Greg: *"There are bursts of it being eight and a half."* (Greg still had his sense of humor.)

Carol: *"Do you know the name of this place where you are right now?"*

Greg: *"No"*

Carol: *"What is today's date?"*

Greg: *"September 13th, 2013."* (The correct date was September 13th, 2011.)

Carol: *"Who is the president?"*

Greg: *"Barack Obama."*

After Greg completed the test for neglect, a perception test given on a large piece of paper, I asked him to *"write your name on the back"*. Greg promptly turned the paper over and wrote "your name on the back." again showing his comedic skills. Funny guy!

I truly believe musicians who have been through trauma recover better when they can play their own personal instruments as soon as possible. Playing them is what their brain has been wired to do, so it becomes a tool for recovery. I always explain to the patient and family that we need to see their instrument as a therapy tool for them to recover and not for performance purposes, at this time. That will come later.

Testing in music therapy. Greg was often monitored for seizures so electrodes remained on his head for some time.

As part of my evaluation with brain injured musicians, I hand them their particular instrument that I purposely place out of tune, then ask them to tune it. Being able to recall their earliest musical training, which is when they learned to tune their own instruments, has become an indication to me that there is hope. If I don't have their type of instrument at MetroHealth, I ask a family member to bring in the patient's personal instrument if they don't mind.

Since we had guitars in music therapy, I handed Greg one that was purposely put out of tune and asked him to tune it. He did it and did it well!

I asked him to name the strings on the guitar. He stated: *"E, A, D, G, B, E."* They were all correct.

Greg was asked to "finger" and play whatever chords he could on the guitar. He correctly played 15 different chords. Amazing!

Months later, after Greg fully recovered, I explained that he had passed the "guitar test" when I first met him, early in his therapy. He did not remember that happening but was proud of himself for being able to do it. I was happy too because then I had no doubt he would recover.

After a few days in music therapy, Sue brought in his music folder that he takes with him to all his gigs. This was a huge, three ring binder with volumes of music inside. Once Greg had that book in his hands, he chose the songs he wanted to play in music therapy. He played and sang them for long periods of time while I sang with him playing a little bit of percussion alongside. Greg was as happy as a clam! He was doing something he truly loved which helped his mood improve tremendously. Sue was so happy to hear him play and sing again. She said, *"this is the Greg I know and love,"* as her eyes welled up with tears.

Greg demonstrated good strength and fine motor skills in his upper extremities as well as memory for playing guitar. It is so fascinating to me that Greg could not remember some 'non-musical' facts; he couldn't remember the current year, the name of the hospital and other details and was slow to respond to questions, but he wasn't slow at playing and singing his songs. He remembered how to play the chords and sing the lyrics. When the music stopped and conversation started, Greg became easily distracted by little things around him. However, when he played his music he could play for long periods of time without distractions.

When I originally asked Greg what he thought his goal should be while in the hospital, thinking he might say memory skills, walking better, etc., he replied, *"to design and build a bridge from the guitar player I was to the person in this place so everyone gets some lovin' out of it."* Even at this critical, confused time in his life, Greg was always looking to bring others happiness. I was touched by this comment and surprised at the depth of its meaning in view of his severe brain injury!

When musicians; instrumentalists or singers, play or sing music after having a brain trauma, their brain is not only doing something it is used to doing but something it enjoys doing. Our favorite music increases the chemical levels in our brains that make us feel good, like dopamine and serotonin, both natural anti-depressants. The reward centers in the brain light up as if the music is an addiction and we want it more and more. Music can reconnect those damaged neuropathways or even make new ones in the brain to facilitate recovery. According to the late Dr. Oliver Sacks,

"Music is the only thing that activates the brain so extensively."

Our goals in music therapy were to:

1. Increase Greg's attention, which was compromised during non-musical activities. As Greg played his music more and more, his attention skills for other things gradually increased. I would play the Djembe, an African drum and sing harmony while he played and sang. This challenged him to focus on what he was supposed to do and when. He actually enjoyed it all.

2. Improve Greg's memory for biographical information, which returned little by little when playing his book of songs. Each song had a particular fact or memory attached to it that Greg would recall and mention in music therapy. He would describe a song he wrote or where he sang a particular song and if there was something special that would happen during the chorus. Music can spark those long term memories, especially when those memories are attached to emotions or moods and piece things back together again like a jig-saw puzzle and then gradually, the short term memory returns. This is a Neurologic Music Therapy (NMT) technique called Music, Mood and Memory Training.

3. Improve his music skills; as Greg practiced his familiar music each day, his music skills returned quickly. It was practically automatic for him.

I often played the Djembe and sang providing an ensemble effect to see if he could continue to focus and attend to his music.

One day, I was on the brain injury floor again, I noticed Greg in Physical Therapy (PT) working on his standing balance. His therapist, Loreen Dobos was timing him as to how long he could stand and not lose his balance. At that time, it was a whopping 13 seconds. I mentioned to Loreen that Greg was used to standing in all his performances while playing guitar. I suggested we have him play his guitar while standing in PT and see if that would improve his time.

The very next day I brought the guitar and Greg's music book to PT. We set his music on an adjustable, bedside table and gave him his guitar. He stood for *15 minutes* as compared to *13 seconds* just the day before, while singing and playing. What a difference! He was getting a lot of attention from those walking by PT. After hearing his wonderful music, they didn't want it to stop because it brought such joy and fun to the whole unit.

Greg was in the rehab center for five weeks. He received music therapy for the last four. Toward the end of his stay, I realized he was improving so much I recommended that he and his two friends who perform together in a group named Horsefeathers, play for our annual MetroHealth Volunteer Luncheon as the entertainment. The Director of Volunteers, Becky Moldaver, agreed it would be a wonderful recovery story to tell! He prepared for this event in music therapy because it was going to take place less than one week after his discharge. Knowing he was going to perform there with two of his best friends was a real motivator for Greg to improve even more. He had no problem remembering the date of the event, October 19, 2011, each time he was asked in music therapy he remembered it well!

Greg was discharged home and then returned to perform at MetroHealth for the annual Volunteer Luncheon along with his lead guitar/bass player, Nick Blasius and singer, Lori Rizzo. Greg was the lead singer and rhythm guitar player. They played for at least 20 minutes and Greg stood up the entire time. He was also interviewed by the TV stations who came to videotape him performing. Greg told his story to different news stations and papers. This truly is a remarkable story and his rehab physician agreed.

Since he left the hospital, Greg and I have kept in touch. He returned to work part time, three weeks after discharge and went full time three months after discharge as the Director of Community and Public Affairs for the Lake County Alcohol, Drug Addiction, and Mental Health Services Board. Although he was not 100 percent recovered, his boss, Kim Fraser was very supportive and patient.

He performs frequently throughout Lake County, Ohio and shares his story with many people. He sings the praises of MetroHealth and music therapy. He is truly blessed to be alive. I have attended some of Horsefeathers' concerts and have had the privilege of singing with them too.

Summer, 2015, Horsefeathers plays at Driftwood Point, Geneva, OH
(Left to right; Nick Blasius, Lori Rizzo and Greg Markell)

"Making music is one of the things I treasure most in life. And I'll forever be grateful to music and music therapy for all the ways they helped me on my long journey back."

Greg Markell

CHAPTER II

Emily
A College Musician Involved in a Crash

"Carol, can you please evaluate Emily Williams for music therapy? She won't open her eyes or do anything. Maybe music will help motivate her, she likes music."

One of the nurses in the Brain Injury Unit saw me walking through to meet with one of my patients. She sounded desperate to help this young, college age girl who lacked motivation to participate in her rehabilitation. She had been in the Brain Injury Unit for four days and had not progressed.

February 25, 2014, six days after her 20th birthday, Emily Williams had been in a horrible car accident. That day there wasn't much snow on the ground but it was -17° in northwestern Ohio. Emily was driving her F150 pickup truck on Route 20 heading for a day at college. As she was driving along this divided highway, she suddenly hit black ice and lost control of her vehicle.

Emily's F150 after her accident

Her truck slid across the grass median, into opposing traffic and in the path of a semi-tractor trailer! Her truck spun around enough so that the passenger side was hit by the semi rather than the driver's side or worse, head on. Thankfully the trailer the semi was hauling was empty. Emily does not remember anything about the accident. There was a state highway patrolman driving behind the semi who witnessed everything. He called for help immediately and an ambulance was there within two minutes.

There had been six prior accidents that day in that same area so an ambulance wasn't far away but unfortunately, her accident was the worst. Her family believes that God orchestrated all these details so her life would be spared; state highway patrolman nearby, ambulance nearby, passenger side of her truck hit rather than head on and an empty trailer. Emily was taken to the nearest hospital. However, because she was unresponsive but still breathing, she was life flighted to MetroHealth Medical Center. The accident happened at 8:30 a.m. and by 9:20 a.m. Emily's mother was called by an emergency nurse at the first hospital, informing her that Emily was on her way to MetroHealth. Emily was wearing her school scrubs with her name tag so the emergency people knew who she was. She was studying to be a Medical Assistant.

Emily's parents arrived at MetroHealth just minutes before the surgeons were going to perform brain surgery because during the accident, a flying object had punctured through her skull and penetrated her brain. Emily opened her eyes one last time and saw her parents. They could tell by looking at her, one on each side of her bed and seeing tears in the corners of her eyes that Emily knew who they were although she could not speak. The nurse inserted an IV into her arm for surgery but Emily was fighting it by moving her arm around. Her father held her arm down so she would not hurt herself or remove the IV. Even under sedation, her parents could see the fight still in her so they were hopeful she would pull through this.

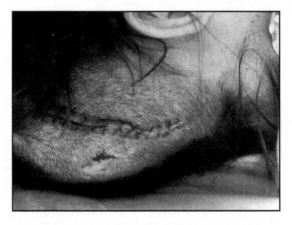

Emily's puncture wound and surgical site at the back of her head

The brain surgery went well; they cleaned out the broken pieces of skull, repaired what they could and placed a titanium plate over the opening they made. The damaged area was in the occipital lobe where her vision is affected and they would have to wait and see what the extent of the damage would be. Emily did not suffer from seizures, swelling of the brain or infections which was a very good thing. She did, however, suffer from headaches and blurred vision enough that she did not want to open her eyes. She did a good job of answering questions, letting everyone know that she had retained her memory and still had the ability to speak. Emily was in the Trauma Intensive Care Unit and Step Down Unit for one week. Given what she went through, this was very good. Even though she did not want to open her eyes, she was evaluated and thought to be a good candidate for MetroHealth's Rehabilitation Center. She was transferred to rehab on Tuesday, March 4, 2014.

Emily became, what seemed to her parents, very depressed. She was crying and not willing to participate in occupational, physical or speech (cognitive) therapies. Her parents didn't know if she was beginning to grasp the extent of her deficits. Her mother, Donna, was not allowed to stay with her overnight for the first night because at that time, this was policy. However, the first night, Emily was able to get herself out of bed and this concerned nursing because she could have fallen and hit her head. She would not cooperate in therapies the next day. When Emily's mother arrived the following day, the nurses saw an improved change in Emily's behavior so they allowed Donna to stay with her even overnight for the duration of her rehab. Wednesday, Thursday and Friday mornings were very discouraging because Emily would not do therapies and would often cry. She would hardly open her eyes. According to Emily's mother, the therapists were trying their best but were at a loss of how to best motivate her.

On Friday morning, as Emily stood still, crying in the parallel bars of physical therapy, Donna hugged her and asked, *"what's wrong Emily, why are you upset?"* She said her head really hurt, she couldn't see and did not want to do anything. Emily's mother spoke with the doctor about this and asked if she could be depressed or if something could be done for the headaches. The doctor started her on medication for the headaches. That very day, I walked into her room to meet her and invite her to participate in music therapy.

When I went into her room, I witnessed a young girl lying on her stomach in bed with her face buried in the pillow. Her long, black hair covered whatever part of her face was exposed. Her mother was sitting in a chair next to her bed. It was very quiet.

In a whispering voice, I introduced myself and explained my role in rehab. I described how music could help Emily in her recovery process. Emily didn't move but her mother proceeded to speak softly to me and described her daughter's background in music. She played piano, clarinet and sang primarily in her church worship band. Her mother was a piano teacher, she taught Emily how to play piano and read music. Donna would often accompany Emily when she sang at church. I didn't think she could speak because she never entered our conversation nor moved until her mother directed a question directly to her regarding what type of music she enjoyed playing. Donna gave examples of genres she thought Emily enjoyed. She moved a bit in the bed and quietly answered *"yes"* to what her mother said but never opened her eyes. I was very surprised she answered the question. I explained that I wanted to take Emily downstairs to complete a music therapy evaluation. So Emily's mother helped her into a wheelchair very slowly and we escorted her downstairs to our piano. During the entire route, which took approximately five minutes, Emily never opened her eyes.

When we entered the music therapy room, I put her at the digital piano while still seated in her wheelchair and asked her to play something for me, anything she could remember. She immediately opened her eyes to see the piano and proceeded to play one of her favorite songs, *River Flows in You* by Yiruma. Emily played this song at home all the time and could play it from memory. This is a beautiful, relaxing piano piece.

At first each hand was playing in a different key. She primarily learned the piece by ear although she knew how to read music. She couldn't remember the left hand part very well so she played slowly and made many mistakes but I could tell she was a pianist. Emily kept struggling through it, for this was *her* passion.

I looked at Donna in another part of the room listening to her daughter. She was happy to see her playing with opened eyes, trying to do something meaningful again even if it wasn't perfect. I told her mother, *"it will come back."* I told Emily that she was talented but would have to work with me in therapy to once again play as she did before her accident. We needed to use music to stimulate her memory. Emily was willing to do that and looked me in the eyes with a glimmer of hope that someone could help her do what she loved to do, and that was to play and sing her music again. Emily kept her eyes open during our ride back to her floor and room.

By Monday morning, all the other therapists couldn't believe the changes. Emily was opening her eyes and anxious to do whatever they asked her to do.

From then on, Emily's attitude began to improve. The only time she closed her eyes was to sleep at night. Emily and her mother really looked forward to going to music therapy so we made it a family affair. Emily wanted to work on her voice and clarinet as well as her piano playing so her mother would accompany her whenever I had her sing. Music therapy helped Emily become motivated to do all her other therapies, improve her memory, respiration and her cognitive skills. Emily also added art therapy to her routine.

Before she was discharged, Emily performed a short concert for all the staff on the brain injury floor by performing one piece each on piano, clarinet and voice with her mother as her accompanist. Emily introduced each piece and could stand while performing on her clarinet and vocal pieces. The piano arrangement she first played for me during her evaluation, *River Flows in You* was now perfect and beautiful in so many ways! She performed it from memory. Emily and her mother were so proud that an important part of her life was once again "normal" and music helped her to recover other non-musical areas of her life, such as; memory for biographical information, coordination skills, social skills and a connection with her family and friends once again.

Emily was discharged from rehabilitation on March 21, 2014, 17 days after admission to rehab. She persisted through her outpatient occupational and cognitive therapies for another eight weeks. She continued to have left hemianopia, which is a decreased vision or blindness in half the visual field. The nerve was injured and there was no definite answer as to how much of this problem might repair itself. Over the next several months Emily had many ophthalmology and rehabilitation doctor appointments. Because of her compromised vision, she was not approved to drive. Her family needed to drive her everywhere she needed to be. This was the biggest adjustment for her, a young, college age woman. Because of her fractured pelvis, blurred vision and headaches, she needed clearance by her orthopedic and rehab doctors to return to work. She was able to return to her job at a sub-sandwich shop, on light duty, ten weeks after her accident. Her left peripheral vision was slowly returning.

Emily went back to college for medical assisting on June 30, 2014, four months after the accident. She had completed seven weeks of school prior to the accident but needed to start all over again forfeiting both money and studies she had completed before. Fortunately for her, the first seven weeks ended up being a review and she did well which boosted her confidence.

Upon returning home after being discharged from rehab, Emily immediately jumped back into all her ministries at the church; Praise Team band, Power Point and Children's Ministries. However, this became too much for her on top of work and school, as a result her tremors and headaches started to return. She needed to give up these ministries temporarily until she completed school.

Her biggest hurdle was being allowed to drive but she needed to wait an entire year before it would be considered.

You see, this was Emily's second car accident. The first accident was due to inexperience and being in a rush. She was 16 ½ years old, having her drivers' license for only four months. It was October 2010 and Emily was in a hurry to get back to her high school for marching band before a football game. She rear-ended someone.

After taking two drivers' tests and drivers' training, she began driving once again on April 15, 2015, one year and 49 days after her devastating accident.

Emily completed her college training on June 12, 2015 and passed two state tests to become a Certified Medical Assistant (CMA). Three days after completing these tests, she started her new career in a doctor's office as a CMA. She has been working full time since then. Emily has done well and her family is very proud of her.

Emily's family at graduation. From left to right: Emily's parents, Tim and Donna Williams, Emily Williams, her sister and brother-in-law, Megan and Daniel Lubinski.

Emily said *"Music therapy showed me I hadn't lost everything. I could still play the instruments I loved and I even still had some of the music memorized. Music has always been my life and even though it felt like I had lost*

everything, I still had my music to help me get through."

Emily's mother added, *"The day I heard Emily play in the hospital, my whole outlook on the situation changed. I knew that Emily was still in there and she could work her way back. Music gave her the motivation she needed to accomplish it!"*

Every time I hear *River Flows in You* or the other pieces Emily played at her discharge concert in the hospital, I think of her. I know she and her family believe God had more plans for her future and she would live to tell her story. God also put music in her life for a reason and one particular day it helped to bring her back!

I wish you well Emily!

CHAPTER III

Richard, A Pipe Organist's Horrific Bike Ride

The organ pipes at St Paul's Lutheran Church, Berea, Ohio

On June 29, 2014, Richard Densmore and his wife Mary had just returned to their home in Strongsville, Ohio after vacationing on Kiawah Island, South Carolina. While Mary and their adult son Alex were unpacking the car, Richard decided to take out his bike, pump up the tires and take it for a quick spin around the block to stretch and check the air pressure. Being the avid bike rider he was, Richard, who otherwise always wore his helmet, decided it would be such a quick trip that he didn't need it nor his wallet which included his identification. Mary and Alex were in the house busy unpacking and didn't see Richard leave but later when they went outside they noticed the bike was gone and the tire pump was left out. They just assumed he went for a quick bike ride because he left his wallet in the house.

On the way home in the car, Mary and Richard had made plans that after all the unpacking was finished, Mary and Alex would walk three miles to a nearby restaurant and Richard would drive the car up later to meet them and have dinner together. When Mary and Alex left on their walk they called Richard on his cell phone to say they had started out. He didn't answer and they had to leave a message.

They didn't know that Richard had already been in a horrible bike accident and no one witnessed it. A woman driving by stopped when she saw a man's body in the road with much blood loss. She couldn't leave her car because she had a young child with her but she did dial 911 to get help. Richard was lying in the road, unconscious and bleeding profusely, with his bike unharmed but a good distance from his body. Two other cars pulled over and the man in the third car got out and went beside Richard to comfort him until the ambulance arrived. With no identification, they didn't know who he was nor whom to call. The ambulance took him to the nearest hospital where he became a "John Doe." The amazing thing was that Mary and Alex had to walk right by the accident scene but by the time they got there, everyone was gone and there was no evidence of any kind. The accident happened at the end of their street and around the corner but neither Mary nor Alex heard the ambulance. When they were about a mile into their walk, Mary received a call from the nearby hospital. The staff noticed the last call on Richard's phone and decided to call that number back. They told Mary that Richard was in a serious accident and was unconscious. They said he needed to be intubated and they were preparing to life flight him to MetroHealth, the Level I Trauma Center (at that time) because he required trauma care. When they realized that Mary and Alex had to walk a mile back home to get the car, they advised them to go straight to MetroHealth.

Richard needed an array of tests, scans and x-rays. He was diagnosed with diffuse bilateral brain hemorrhages, frontal brain contusions, bilateral orbital fractures, a left rib fracture, multiple facial fractures, left gluteus hematoma, a collapsed lung and bilateral temporal bone fractures. He needed a chest tube placement and a bolt was inserted into his brain which monitored his intracranial pressure. He had a nasogastric tube inserted and spinal cord fluid was leaking from his nose. He was in very bad shape and non-responsive.

Richard was sixty-years-old and needed to be admitted to the Intensive Care Unit (ICU) where he stayed for two and a half weeks. What a traumatic time for the family. Mary and Richard have wonderful relatives who were all praying, helping and supporting in any way they could. The family did not know his prognosis.

During his time in ICU, Mary told the staff about his job as a librarian at Baldwin Wallace University and as the Assistant Pipe Organist at St. Paul's Lutheran Church in Berea, Ohio. The family would play the Classical music cable station on TV for Richard to hear. One of the male, night nurses made a playlist of Classical music on his personal iPod for Richard to listen to. How thoughtful was that? The family would also take turns reading Jane Austen's *Persuasion* to him. Jane Austen is his favorite author. Richard regained consciousness on July 4, was extubated on July 5, 2014 and received extensive plastic surgery on July 11.

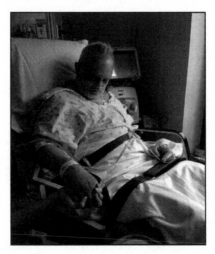

Richard in ICU

Eventually he was transferred to the trauma Step Down Unit for a while. The staff needed to decide where to discharge him; MetroHealth's Rehabilitation Institute of Ohio or a skilled nursing center. He was not able to go home yet, not even close. There was great concern whether Richard could handle the extensive therapies he would undergo if admitted to MetroHealth's rehab center since Richard was extremely fatigued. Even his neurologist told Mary that they consider anyone over the age of forty to have an *old* brain. (Wow! I'm in trouble!) After evaluating Richard and discussing options with Mary, the decision was made to transfer him to MetroHealth's rehab center. When he was admitted, his physiatrist immediately referred him to music therapy upon learning of Richard's love for music and his professional pipe organ position.

I met Richard and Mary almost one month after his accident on Friday, July 25, 2014 at his music therapy evaluation. He was so fatigued after having three hours of his other therapies that I had to split his evaluation over two days, 30 minutes each. He was restless and sore from his other therapies. His affect was flat which is typical with brain injury or stroke. At times, he repeated what I said, rather than answer my questions, which is a form of echolalia. He was only oriented to himself and could not identify the date, time, place or why he was in the hospital and exhibited decreased attention and concentration. Richard's speech was slow, soft and he used short phrases or one word answers when they were appropriate. I measured his respiration to be 7.25 seconds, meaning the length of time he could sing one note on a single syllable like "ma" and this was very poor. His vision was very blurry and his left arm and leg could barely move. His injury caused stroke-like symptoms.

Mary provided Richard's musical history for me to complete his evaluation. He was hired as the Assistant Church Organist at St. Paul's on a part-time basis. His full-time career was Librarian, Head of Public Services at Baldwin Wallace University. Baldwin Wallace was also where Richard took organ lessons during his college years and at one time was an organ major.

He also sang in the church choir at St. Paul's as well as the Oberlin College's Musical Union Choir, a community based oratorio society, a hobby he and Alex shared.

Richard grew up with music as he learned to play trumpet for his high school band, started taking piano lessons at the age of 16 and then took organ lessons when he entered college. Music was frequently played in his home as a young boy. Later in his recovery process Richard was able to share with me his memories of singing camp songs while traveling during the summer with Don, his dad who was a truck driver. Don was also a square dance caller. Music was extremely important to Richard while growing up and is still today. He is very proud of owning not one but two Edison Diamond Disc phonographs, frequently listens to his collectables and loves to share information about them with whomever walks into their home. Under the heading of "lesson learned." be careful of asking him questions about the Edisons for he provides many volumes of detail.

Now, in rehabilitation, he needed to use music as a therapy tool to heal his brain, improve his motor skills and bring him back to the person he once was before. Music would help fire and synchronize the brain neurons needed to promote muscle movement, generate memory and cognitive skills as well as stimulate appropriate speech. As I've described in prior chapters, music can do these things because it activates the brain so extensively and all at one time. Music helps to organize the neuropathways that have been compromised from injuries.

Both sides of Mary and Richard's family were extremely supportive and helpful during this time. Someone was always at his bedside or with Mary helping her keep up with the everyday chores of running a household. She also worked for a church part-time but they were very flexible and understanding with her hours during Richard's recovery. A good support system helps promote a good recovery.

As I proceeded to evaluate Richard, he could actually identify all the notes on the keyboard but had difficulty reading music due to his limited vision. His left arm was flaccid and the fingers on his left hand moved very little because they were extremely weak. His left hand could not even hold a small, hand held percussion instrument like an egg shaker. His right hand could start a melody on the piano only after I provided a musical cue. For instance, I would start to sing a song familiar to him and he would begin to play the melody but couldn't finish it and would make something up. Richard was used to reading music and didn't play very much from memory or by ear.

I encouraged the family to bring in from home a CD player for his room and his favorite CDs. I explained that if he listened to music every day and sang to his favorite songs, especially with family, this would start to heal his brain, his lungs, start to bring back memories and improve his speech. I explained this was something they could do together when he was not in music therapy, during the evenings or weekends.

They followed through with my recommendations and enjoyed doing it together as a family. It gave them all something very positive to do with and for him.

In music therapy, over the course of three and a half weeks in rehab, Richard showed improvement every day. Our goals were the following:

Improve his speech using an NMT technique called Music Speech Stimulation. This technique requires the patient to complete phrases from familiar songs. I would start to sing his favorites but in each song, would leave out phrases that he would have to complete. Richard did this very well. It was easy for him.

Our second goal was to improve his left-hand grasp. A technique called Therapeutic Instrumental Music Performance uses instruments in strategic places or in a variety of ways to increase upper extremity strength, range of motion, dexterity and fine motor skills. We used this technique while listening to music with a strong rhythm to assist with his left-hand grasp. At first, I literally needed to hold Richard's left arm up and put my hand over his hand to hold an instrument. He couldn't hold anything with his left hand. I also needed to hold up his left arm while he started to learn piano again even though his fingers could barely push the keys. This was important for him to do. Electronic keys are easier to depress than keys on an acoustic piano. That is where we started because a slow, steady approach was needed. We used a metronome during scales and finger exercises for strengthening and dexterity. As I've said before, a perfect, regular rhythm helps fire motor neurons in the brain to aid in muscle control.

Our third goal was to improve his respiration by using a different technique called Therapeutic Singing. Singing demands more oxygen intake, helping to cleanse and exercise the lungs and further assists in preventing pneumonia. During music therapy his family often loved to sing with us using Richard's favorite hymns and camp songs. Richard enjoyed singing.

Goal number four: Organization of thoughts; I used what is known as Music Executive Function Training. His thought processes became more organized by re-learning to play piano. This takes a multitude of skills at one time; planning ahead, reading music and expressing that symbolic reading system into motor skills. In addition, I gave Richard a musical challenge; categorizing genres of music which I provided for him. When I played different genres his task was to identify whether it was Country, Classical, Rock 'n' Roll, etc.

His wife said he was always so motivated to attend music therapy and much more verbal after music therapy than any other therapy.

By the time Richard was discharged from rehab three and a half weeks later, he could hold and play all seven hand held percussion instruments without help from me, using his left hand. He was also able to independently play three simple songs on the keyboard, reading music from a level one piano book with both hands simultaneously. The book used very large notes on an elementary level. This was still an accomplishment because of his blurred vision. Richard not only saw double lines of music but one of the lines was crooked to his eyes. His respiration greatly improved from 7.25 seconds to 21 seconds. His memory started to return very well and his echolalic speech was resolved because he was answering all questions appropriately. This was all accomplished by August 20, 2014 when Richard was discharged.

The family was so very impressed with his rapid improvement. They knew that music was his love and desired for him to return to his church as the Assistant Organist so they hired me to provide personal music therapy services in his home after discharge.

I provided two, one hour sessions per week starting after Labor Day, 2014. By Christmas he could play every song in the *Family Book of Christmas Songs and Stories* by Jim Charlton and Jason Shulman, 1986. These fifty songs were at a moderate skill level and were smaller print than what he had in the hospital. He even performed some of them for family during the holidays. Our new goals were to continue to improve strength and flexibility in his left hand and to further improve his music reading abilities. We needed to work on those goals first before introducing the organ with foot pedals which would be our next task.

Richard's family was so happy to hear him play again, even if it was the piano at this point and not the organ. I received a very sweet note from his sister Jane, that read something like this; "*Carol, thank you for bringing our brother back to the person he was.*" The note also came with a wonderful box of chocolates! Now that was "sweet."

Richard was still going to outpatient therapy at MetroHealth working on balance, gait, left side shoulder and upper extremity range of motion. He needed physical, occupational and speech (cognitive) therapies. He still needed to use a walker and gradually improved to the use of a cane. His vision remained blurry but was improving.

It was evident that he was not ready to go back to work and needed to go on disability. A librarian's job is meticulous. He would need to keep detailed records and be able to see well. It required a tremendous amount of endurance which he did not yet have. The family was contemplating his possible early retirement but wanted to weigh it all out. Mary had her part-time job and the family and friends provided donations for all his medical expenses not covered by insurance. Obviously, they needed to watch their finances more closely than usual.

After the holidays were over, I believe the plan was to cut down on his music therapy sessions even though he loved them. Fortunately, Richard's family came through with a wonderful Christmas present. They paid for six months of music therapy sessions with me to continue twice per week. (I should have put a bow on my head!) Richard and Mary were so grateful and I was happy for him too because this meant we could now continue our work on the organ itself!

We eventually decided it was time to use the pipe organ in therapy. I asked Richard to make arrangements with his church for us to have one therapy session at the church each week so he could use the organ. Having Richard make the arrangements allowed him to take some responsibility and he did it well. Our other therapy session each week would continue at his house on the piano. We needed to start using the organ pedals or his feet would forget what to do! *"Use it or lose it"* is a very true statement when it comes to the brain.

Tuesday evenings we were at the church and Fridays at his house. Richard chose typical liturgical, organ music a bit on the simpler side that I would have him play at a slower tempo. He chose composers like Bach, Couperin, Dunstable, Handel, Telemann and others. This was the first time he played the pipe organ after his injury and he was amazed he could still play! What a moment for him! The sound of the pipe organ was very moving for both of us!

Also, just like we did with his hands on his home piano when these private sessions first started, I had Richard play scales to a metronome using his feet on the pedals. Both his feet needed to complete each scale from the lowest pedal to the highest. We started with a slow tempo and gradually increased it. Remember his left side was still weak and slower than the right. This technique helped improve strength and flexibility in his left leg and improved his entire body coordination on and off the organ. I required him to use the metronome with all his pieces because his tempo needed consistency. I know Richard hated the metronome. I think he once told Mary over dinner that he wished I would lose it. I told him to think of it as his friend but I know when its batteries died a couple of times, Richard was a happy camper. Cruel therapist that I am, I downloaded a metronome app on my new (and very first) smart phone. Good-bye happy camper!

Practice, practice, practice!

From January to June, 2015 we continued with two sessions per week with one each week on the pipe organ. By June he could play half the music for a church service but he also played for two back to back services one Sunday. This was now one year after his accident. His pastor announced to the church at Richard's first Sunday back, *"when I first saw Richard in the hospital, I thought he would never play the organ again."*

I couldn't believe all the music that is required in the Lutheran Church service. Almost one full hour! There is a prelude, seven liturgical pieces, six hymns each with at least three or four verses, a special choir piece to accompany and a postlude. Think about that; he had to read music, use both upper and lower extremities for almost one hour with just a few very short breaks. He needed to be organized and prepared with each piece, knowing when to play AND what combinations of stops, which open or close different sets of pipes for each piece. This requires enormous, physical and mental endurance, strength and flexibility, planning and executive functioning as well as musical expression.

We thought the goal for him to play for just part of the service was good for the summer months. As Assistant Organist, he only played once per month and not each Sunday even before his injury. During the summer months the church has two services back to back each week. Needless to say, his endurance was challenged. He was tired after the first partial Sunday he played. In July he was on a family vacation and then in August he once again played the other half of two services.

By September of 2015 he was back to playing a full service, one service per month. I attended that service and if I have to say so myself, he did a fantastic job! People were happy to see him officially back again.

At this time, he was volunteering his services because he saw this as part of his "therapy." Since Richard wanted to do well and was dedicated to doing this, he would show up at church early and practice about thirty minutes before the service started. He was a tired puppy when he went home!

At this point in the autumn of 2015, instead of eight therapy sessions per month we cut down to seven. Richard used that extra day to practice for two hours on Friday mornings (before his Sunday at the church) by himself. We were still working on good practice procedures, technique, endurance, planning and preparation before beginning a piece.

The executive functioning skills are the highest level of cognitive abilities; decision making, organization of thoughts, planning ahead, etc. and seem to be the last abilities to return after a stroke or brain injury. Richard would choose a piece of music and just start playing it without looking at its details or preparing his mind. He would also play through his music during his practices, in-between my sessions and not often work out difficult places. I guess some people may not call it therapy at this point in his recovery but more like organ lessons. Although I am not an organist, I do know the musical techniques for proper hand positioning, playing, dexterity exercises for his hands and feet as well as mental planning exercises in music. I do know whether or not his tempo is consistent and whether or not he is playing the piece properly. I make recommendations for improvement based on my therapeutic experience and musical background and Richard loves to teach me all about how the organ and pipes work! Just ask him a question about the organ and his eyes light up. Of course in his mind, pipe organs are the only REAL organs.

As of January 2016 Richard and I still worked together but at that point we met only six times per month. He seemed to enjoy the challenge I provided, the consistency in his playing abilities and the accountability he had to me rather than just practicing on his own. I believe when we have to report to someone and show progress each week, we work harder than if we didn't have to be responsible and just practice without a therapist or teacher supervising. That's our human nature. Accountability makes us work harder and/or smarter.

Richard was now being paid as the Assistant Church Organist once again and loved doing it.

As of July 2016, Richard and I met for his lessons once per week, down from six to four times per month.

Like many trauma victims, Richard doesn't remember the accident or what happened. I believe that's the brain's way of protecting us from having nightmares. The police hypothesized that perhaps a car cut him off while riding his bicycle or an animal jumped out in front of him. He may have squeezed the brakes too hard, stopping the bike abruptly, catapulting his body into the air and then down onto the pavement. The bike was unharmed so they don't think a car hit him. Unfortunately, Richard doesn't even remember the wonderful vacation they experienced just prior to his accident. Maybe those memories were so new they didn't yet transfer from the short- term to the long-term memory centers in his brain.

Richard started driving one year after his accident, started mowing the lawn in the summer of 2015 and was back to being in charge of the laundry. (Mary told him that before his head injury he used to love doing the laundry!) He was even back to playing golf, swimming and hiking. He actually went on roller coaster rides at Cedar Point in Sandusky, Ohio one year after his accident at the age of sixty-one! During the winter of 2015 Richard was so proud of being able to work the snow blower by himself.

Because I have spent so much time with Richard and Mary for more than two years, they have become like family to me. We often share stories of our children, siblings, work or politics!

In Richard's words *"Music therapy was important to me because I knew there was someone who could see my potential."*

After his recovery, Richard proudly shows his two antique Edison Diamond Disc phonographs behind him and holds an actual disc.

CHAPTER IV

Daniel, the Piano Man
A Hit and Run Bike Accident

Daniel Gendrich, Circa 1991

June 4, 2011; having had a little too much to drink because he and his girlfriend split up, Daniel took off on his bicycle without wearing a helmet and was suddenly hit by a car.

The exact details may never be known because the driver left the scene of the accident and Daniel was found on the side of the road, unresponsive. When he arrived at the hospital, he was classified as a level II trauma patient. The tests revealed he suffered blood clots on both sides of his brain as well as brain hemorrhages, cheek and temporal bone fractures along with a collapsed, right lung. Daniel had many lung and brain contusions. He started to experience cerebral edema; the increase of fluid in certain parts of his brain and his contusions were increasing while in the ICU. Daniel became so confused and agitated that he tried to pull out all the tubes used for his care. He had no idea where he was.

By the time I met Daniel, June 10, 2011, he was admitted to the Rehabilitation Department - Brain Injury Unit after receiving six days of combined intensive care and step down care. Even in view of his extensive injuries, he didn't take long to improve to the point where he could be transferred to Rehab. Although a bit confused, this was amazing considering his injuries.

Once in Rehab, he began to realize he was in a hospital. He remembered his full name, "*Daniel Gendrich*" and the current month but didn't know his birthday, the current day, date, year or name of the hospital. In addition to his memory loss, he was experiencing a very mild form of aphasia, the inability to express himself correctly. He would sometimes substitute words for what he was trying to say. For instance, when I asked Daniel why he was in the hospital, he stated, "*I was driving a car and fell.*" He meant to say a car hit him and he fell off his bike.

During his first music therapy session, Daniel expressed no affect, hardly established eye contact with me and was impulsive; everything he said and did was done quickly. These are all symptoms of brain injury; not unusual for this part of his recovery.

Daniel clearly remembered he was a piano player! That was interesting to me because there was so much he couldn't remember. However, he could not identify musical notes on the music staff nor on the piano itself. "*I don't know why I can't figure this out,*" he would say while looking at notes on the staff. "*I don't know why I can't remember that,*" when I asked him to identify the note B on the piano.

In addition to his impulsivity, his attention was impaired. He had difficulties concentrating and was slow to process information.

He knew that he loved to play piano for parties, night clubs, churches and bands. He also remembered that he liked Classical, Jazz, Rock 'n' Roll, traditional church hymns, show tunes and easy listening music. Before the accident, he often played in the evenings after he was finished with his day job, selling car parts for Euclid Foreign Motors.

When I asked Daniel to play something for me during his first session, he sat at the baby grand and tried. It sounded somewhat disorganized, I wasn't sure what it was but it was evident he was a pianist. However, as he persisted he realized it wasn't coming together. He couldn't continue or remember how to finish a song. "*I had a bad accident,*" he said, providing an explanation for why his abilities were compromised. I asked him if he knew how to play particular songs but I would have to start singing them, providing auditory cues to help him recall the melody. He would begin with the song I was singing but when I stopped singing, hoping he would recall the rest of the song, he then reverted to what he was originally playing. Every song he played eventually sounded like the same tune. As a music therapist, I could hear that he had the skills of a pianist although his method was fractured and disorganized. As a result, we had to start all over again using primary piano exercises. He had to learn to read music from the beginning and take it slowly, learning notes in the treble clef for his right hand and bass clef for his left. He slowly graduated to more complex music.

Daniel had no physical impairments. As a matter of fact, by June 14, his first day in music therapy, he walked with me down long hallways from the eighth floor to the seventh floor using an elevator only to avoid the stairs. His problems were his memory, attention and lack of organized thoughts. I was impressed with his physical abilities this early on in his recovery.

He was divorced but his ex-wife and her husband would visit him in the hospital. They would visit him frequently and they were the people who helped me understand some of Daniel's musical history. This very concerned and helpful couple told me that he once played sousaphone in college and had the honor of "*dotting the i*" in The Ohio State Marching Band.

At this point in Daniel's story, I think it is imperative to understand the extent of his musical background and early training. These dynamics play the leading role in my assertion that the brain of a musician recovers more efficiently after trauma, than others when using music therapy techniques. That said, Daniel began his formal music training on the accordion at the age of 6 ½ and for the next 15 years, studied under the guidance of Walter Gaus, *arguably one of the best accordion players in the world.*

At the age of 10, he added a second instrument, the piano. Piano lessons continued through his high school years and beyond. Also, during his years at St. Joseph's High, Daniel taught himself how to play a third instrument, the sousaphone - for the marching band.

Around age 13, Daniel entered a state-wide Accordion Competition held at Cedar Point and won second place!

Daniel was accepted not only to The Ohio State University but was also accepted as a member of their prestigious, marching band. He was so impressive that he was given one of the band's highest honors. He and his sousaphone would be the "dot on the i" when the band spelled out the word Ohio on the football field. For those who don't know, just being accepted into the OSU marching band is one of the most difficult achievements but to be the "dot" is beyond incredible! During Daniel's tenure at OSU, the marching band played for President Nixon's inauguration as well as three Rose Bowls! What joy music brings.

After receiving a dual degree from OSU in piano performance and music education, he continued his piano studies at the Cleveland Institute of Music, another difficult school in which to be accepted. Sadly, for personal reasons, he couldn't complete his studies there.

Daniel eventually told me that his idol was Liberace! He always wanted to play like the ever so fabulous Walter (Wladziu Valentino) Liberace.

While still in the hospital, Daniel seemed very disorganized, even his clothes looked disheveled. He was confused but he knew exactly when music therapy was scheduled. Each patient room included a white board installed on the wall where their daily therapy schedule was written. Most patients in his condition didn't pay attention to it but Daniel did. He often gave other therapists a difficult time about going to their therapies but not me. The minute I walked into his room to take him to music therapy he not only knew who I was but immediately stood up, picked up the brief case of music, which he asked his son to bring in and started walking to the music room. This is the room with the baby grand piano, similar to Liberace's only not as fancy and certainly no candelabra!

In music therapy Daniel had to study the notes on the sheet music as well as the keys on the piano itself. He attended music therapy five days per week from 40 to 60 minutes each day. Three days after his evaluation Daniel could identify 30 percent of the keys on the piano with no cues. The other 70 percent still needed cues from me. Using sheet music, he could identify all the notes in the treble clef with minimal cueing. The notes in the bass clef were much more difficult for him to remember. However, Daniel could sight read a very basic, primary song using both hands independently, reading treble and bass clef. This motivated him to keep going. By now, only three days after the evaluation, Daniel could play many songs after hearing the music first but could not remember the song by its title alone. His musical ear was quickly returning but he still needed auditory cues. He was progressing quickly and continued to work very hard. Every day Daniel would remember more and more about the piano and his music reading skills. Each day he sounded better, playing with improved technique. After his music therapy evaluation, he received a total of only six sessions.

His friends recommended calling the TV stations to tell his story. They thought that since he once *"dotted the i"* in The OSU marching band and now re-learned how to play piano, this would make a good story, so we did. One of the local TV stations interviewed Daniel so he could tell his personal story about his accident, brain injury and journey through music therapy. The reporter also interviewed his doctor about the type of brain injury he sustained and they interviewed me as well about Daniel's time in music therapy.

The day before he was discharged on June 24, 2011, he provided the reporters, family, friends and staff with a mini concert using some of the old standards. The camera was rolling while he played:

Fur Elise by Beethoven

What a Wonderful World by Bob Thiele and George David Weiss

All of Me (the old jazz tune) by Seymour Simons and Gerald Marks

Georgia On My Mind by Hoagy Carmichael and Stuart Gorrell

Don't Blame Me by Jimmy McHugh and Dorothy Fields

Don't Take Your Love from Me by Henry Nemo

Somewhere Over the Rainbow by Harold Arlen and E.Y. Harburg

My intern at the time, Cheryl Pasquale, created a "set list" for him to refer, to help keep him on track with the selections. He performed very well and his interview and concert aired on the news that night!

Several months later, MetroHealth's Foundation had their ever so important, annual 'Donors Thank You Reception.' This would be a large group of physicians, administrators, business people of the community, stakeholders and other staff. They asked me to provide the entertainment for the reception using former, recovered patients who are musicians. What a wonderful way to showcase the benefits of their donations by using their money toward successful therapies, techniques, instruments and equipment to help patients reclaim what they love to do.

Like Greg Markell in chapter I, Daniel was one of several I invited back. The funny thing was, I worried about whether he would dress appropriately for the occasion. Well, did Daniel surprise me! He not only showed up in a beautiful tux but wore a sparkly bow tie, with lots of bling as well. *"Carol, guess where I got this bow tie?"* I responded, *"where?"* He said *"I bought it at a Liberace auction."* We both laughed loudly, he wanted to play **and** look like Liberace and yes, it actually belonged to his idol! Daniel sounded wonderful, playing Classical music selections and popular show tunes. His fingers went up and down the keyboard with ease and beauty. I watched in awe how he played with confidence and passion. The only things that were missing were the candelabra and diamonds.

Liberace's actual bowtie!

About a year and a half later, the entire Rehabilitation Institute of Ohio moved out of the main hospital facility and into the Old Brooklyn facility of MetroHealth, one and a half miles down the road. We celebrated the opening of our new home with food, speeches and of course, music performed by musicians who were once patients. Again, Greg and Daniel were two of those musicians I invited to return. Lori Rizzo harmonized beautifully along with Greg as they usually did for their gigs. Daniel played up a storm on the digital piano with his old but memorable brief case of music beside him. When I looked at the briefcase, I remembered how hard he worked and how dedicated he was to his music. He played elegantly!

Daniel retired from his day job as a salesman but continued to play professionally. It didn't take him long to get back to "normal". He was so happy, looked well and gave back to the hospital and department that helped him return to the life he loved.

Daniel recently told me that after retirement, he went back to work at Euclid Foreign Motors as a salesman two days per week and proclaimed *"my (music) skills have not deteriorated one bit."*

Instead of carrying that heavy, old briefcase, Daniel has 'updated' himself by scanning all his music selections to his iPad and reads his sheet music that way while playing. He proudly told me that he was preparing a full piano concert for the Lithuanian Club where he planned to play selections like *The Grieg Piano Concerto in A minor and Piano Concerto #1 in B flat minor* by Tchaikovsky along with other famous pieces and electronically adds other orchestration.

Further, at Thanksgiving, 2016, the 53rd annual "Polka Party Weekend" took place. This four-day event is a World Wide Accordion Jam Session

and was held at the Marriott Hotel, downtown Cleveland, Ohio. Daniel performed with the best of the best in the world and had the time of his life!

"All the people at MetroHealth were friendly and nice. I would recommend Metro to anybody. Music therapy got me back to doing what I do best and that's playing the piano!"

Daniel Gendrich

CHAPTER V

Chad, A Guitarist's Four Story Fall

Chad's mother, Teri Miller, remembered hearing her cell phone ring around 11:45 p.m., St. Patrick's Day, 2016. She was sleeping. Initially startled, she decided to ignore the Pennsylvania phone number unfamiliar to her. Just another bogus call…. no message. It wasn't until the same number reappeared on her screen seconds later that she began to take notice and answered.

That moment would forever change all their lives.

"Mrs. Miller, this is Joey, a friend of Chad's." "Yes?" She answered with dread. *"Chad was leaning against a railing and fell. He was taken to the hospital and his vital signs are stable."* She felt her heart rate speed up. She tried to ask more questions, searched for understanding, tried to stay calm but couldn't so she woke up her husband, Larry.

They looked up the phone number for Kings County Hospital in Brooklyn, New York since Joey informed them where the ambulance had taken Chad. After what must have seemed to be a life-time on hold and multiple transfers, the Emergency Department (ED) resident, confirmed their worst fear. Chad had fallen four stories, was unresponsive and "very critical." The ED resident's last words were *"get to King's County as soon as possible."* They had no idea if they would ever see their son alive again.

The trauma team evaluated him; he was intubated and the CAT scan of his head confirmed he had a subarachnoid hemorrhage, a subdural hematoma and an intraparenchymal hemorrhage with midline shift. His heart rate dropped dangerously low so he was given medication to help bring it up. He was extremely critical.

His mother remembered asking if the hospital was a trauma center. As a nurse practitioner herself, she knew head injuries would require this. Yes, it was a trauma hospital. Trauma centers are equipped to care for patients with multiple, traumatic injuries. Otherwise, survival, if at all possible, would be unlikely.

So many things to think of all at once: responsibilities, immediate arrangements needed to be made before leaving so Teri called her mother crying. She was calm, prayerful, always the matriarch. She would call the family, pray unceasingly, take care of their dog Lilly and their home.

Teri was already somewhat packed because in two days they had planned to visit Chad for the first time in Brooklyn, since he moved there only 10 weeks prior. They were very much looking forward to a fun trip, finally getting to see Brooklyn and the rental he found on Craigslist. He was sharing a four-bedroom house with three other guys. It was going to be a weekend of celebration.

Larry was ready within minutes. They left their home near Cleveland devastated and frightened. They contacted both of their daughters, each at separate colleges in Ohio. Larry and Teri made the decision to drive to Brooklyn since waiting for an early morning flight was simply out of the question. After gathering both daughters, each about 45 minutes in opposite directions, they hit the road. Initially, they struggled with the decision to take the additional time but in the end, they knew they needed to be together. *"Our only security was knowing that whatever the outcome, our family of five would be with one another, perhaps for the last time."*

The overnight drive to Brooklyn was a mix of silence and tears. Larry drove with a passion to get to New York in record time. During the trip, they received a phone call from Kings County, the neurosurgical team wanted permission to perform a craniectomy. It would be Chad's only chance at survival since the head bleeding was producing life threatening pressure on his brain. Removing a piece of the skull would help relieve the compression on the brain or death would likely ensue. This was clearly emergency surgery which would be performed by a neurosurgeon who was on call that night.

Teri said *"yes"* not knowing the reputation of the hospital nor the surgeon and knowing that they'd be driving the entire time, she couldn't be available to Chad except through prayer and pray they did! The operating room was ready and the surgery would likely take one to two hours. She later recalled that she granted permission to them by completely surrendering to God, devastated, pleading for his life to be spared, hoping they would see him alive.

The family's Pastor, Dr. Dave Collings, over the years urged her to memorize scripture. Recently, she was given a book called Psalm 91 by Peggy Joyce Ruth. This scripture was one of her favorites, she memorized this psalm because she prayed it daily asking for protection over her family. For many reasons she fretted over their safety, their decisions and their absence from her home. She prayed this Psalm out loud many times during that long ride to New York. It was etched into her heart. Little did she know how many times this prayer would get her through the coming weeks and months.

They arrived in Brooklyn after about eight hours only to get caught up in the morning rush hour. It took close to two hours of stubborn NYC traffic to reach the hospital. Exhaustion, dread and panic consumed them. When they finally arrived at the information desk, hospital policy dictated that all be photographed and produce ID before being allowed admittance to the ICU; yet another delay. They were told to wait in a family area that was not very private.

The senior surgical resident in the ICU introduced himself and gave the details of what transpired over the last 12 hours. Chad made it through surgery but was in an induced coma. There was damage to the brain and it was unclear what function remained. This continued to be a life-threatening event, the outcome still unknown.

There were three different teams caring for him; the neurosurgical, the trauma and the ICU teams. Larry and Teri would ultimately meet dozens of health care professionals in the weeks ahead, all of whom had significance in Chad's outcome. All the teams worked together. After surgery, Chad was transferred to the Surgical Intensive Care Unit. The senior, surgical resident led them into Chad's room.

Teri had been a nurse for more than 30 years. As a young nurse, she worked in ICUs. As a seasoned nurse, she visited countless ICU patients as a practitioner for a consulting service. The constant monitoring, alarms, endless technology, energy and competence of the staff and team approach were imperative to save the life of this critically injured patient. Always stimulating, nothing however, can prepare you for the despair gnawing at your soul as you look upon your own child in the ICU, wondering if he will live or ever be whole again. They didn't know if they had only minutes left together or if his life would continue in a coma. As a mother myself, I could fully understand their anxiety; would he ever wake up? Would he ever be "their" Chad again? Praying continually and begging for Chad's life, she felt an eerie sense of calm. The ventilator was breathing for him and his vital signs were stable. His skin was warm and they could finally touch and hold him. Teri said they felt the presence of God Almighty, somehow knowing he was listening.

Their family was all together, however messy and disheveled. Like most families, they had never been in this type of situation before. Larry remained strong and faithful, embracing Chad and telling him how much he loved him. One sister, Danielle, wanted to stay in the unit every minute. The other, Hannah, wanted to be close but felt physically overwhelmed by the sights and sounds. Both reactions understandable, they prayed through their tears.

Thankfully, in walked Joey and he brought his parents, who traveled from Philadelphia. Joey, a friend from college, was Chad's only physical advocate during the entire night of surgery. He was physically shaken but greatly relieved by Chad's survival and the presence of Chad's parents. Joey was the 911 caller, the witness and friend who remained with Chad. Later, the doctors asked how Joey was doing, they were afraid he was in shock. This was so much for a friend to handle alone.

Teri called the manager of Chad's employer. They would have no idea what happened otherwise. Everyone was very concerned and wanted to know the details. Could they visit? Even though Chad had only started there fewer than three months earlier, they all connected like one big family. The owner later conveyed how she came to hire Chad so soon after he arrived in Brooklyn: Chad worked for a friend of hers in Washington, DC and was highly recommended. Her brother had just opened a deli a few blocks from the hospital. Even though he never met Chad, within an hour of her phone conversation, he walked in with bags of food for the whole family. Teri said, *"the meal was plentiful and delicious. We cried and hugged him. I could not remember when we'd eaten last."*

The first day, they were able to meet the neurosurgeon who was on-call the night of Chad's accident and the one who performed the life-saving surgery. Teri said that she appreciated the fact that the surgeon was warm, personable and spent a lot of time with them. However brilliant and conversant, she did not mince words. *"The next 48 to 72 hours are critical if Chad is going to survive."* The chances were high he may not. His brain was damaged; the extent was yet unknown. He would require total life support for now. Long-term ramifications could not be determined. The neurosurgeon asked many questions about Chad as a person and about the family. At the end of the discussion she asked for permission to take Chad back to the OR that night to place a ventricular drain in his brain. This would allow monitoring of the intracranial pressure (ICP), vital information on which to base future clinical decisions. The surgery was scheduled for 2 a.m. They were overwhelmed again. Teri said; *"What a blessing it was that she was on call the previous night. God chose the right surgeon to be with Chad."*

He arrived back to the ICU much later. There was a tube sticking out from his skull attached to a drainage bag containing light cherry colored, cerebral spinal fluid next to a device that measured the critical pressures in his head. This was another line of data on the bedside monitor alongside heart rate, blood pressure, oxygenation, core temperature; constant surveillance. The procedure went well. One moment at a time was all they were given.

Teri was becoming a fixture in the ICU, only leaving for moments to introduce herself to the dozens of friends who would soon make their way to his bedside. The Millers called their families in Cleveland and California; they were emotional, tearful conversations. She soon felt overwhelmed and couldn't handle the influx of texts asking for information so she decided to use a closed blogging site called CaringBridge to keep everyone informed. She could post pictures and daily updates. This site became her companion, her sounding board, the entrance to the soul and exit from the abyss. Word was spreading and she received comments from people she hadn't seen in decades. It became her daily communication life-line. Teri told me that *"God continued to hear and answer our pleas. It felt natural to pray over Chad."*

She thought back to her inquiry that first night when she asked whether Kings County was a trauma center. In fact, it was the first trauma center in the United States. If there was one thing they knew, it was trauma.

Larry and Teri were only familiar with one major Level I trauma center in Cleveland; MetroHealth Medical Center. They really wanted to arrange a Metro Life Flight transfer. They missed their support system but were totally grateful for the tremendous care Chad received at Kings County. On the other hand, Chad's support system was in NYC. The very private Chad was taking on a new identity for his parents with each of his friends sharing what they loved best about him. They heard stories of how he helped them through rough spots and how loyal he had been. His family laughed and cried at their honesty and their encouragement. Support systems are so important; they were elevated daily by these friends and their families.

Teri consulted with a dear friend and former colleague, whose husband worked for Life Flight and the emergency department at MetroHealth. The price tag would be overwhelming to get him back to Cleveland, however, the instability of his situation was, of-course, their bigger concern. Their hearts sank. Would they ever be back in Cleveland again? As much as they were grateful for the care in Brooklyn, when it was safe to move him, that's where they needed him to go. It was clear, however, this wasn't the time.

It had been more than a week in the ICU and Chad was still in an induced coma. Heavy sedation was needed to keep him calm, rest his brain and prevent him from pulling out all his tubes in a state of agitation. Typically in the morning, the sedation is weaned off temporarily to fully assess his neurological responsiveness. Pain brought movement to his right arm and leg but his left side remained paralyzed. His eyes were slightly opened but with no meaningful tracking. There was no squeezing of the hand or signs of purposeful activity. Daily prayers and pleading with God along with CaringBridge filled with messages of support and love kept his family going. Teri told me that when she reached back into the entries on the website, she was reminded that in Chad's recovery, *music was instrumental from the beginning.*

His friends played music for him from their playlists with selections they thought he would like. His mentor and friend brought a playlist that Chad created of his top 10 songs along with his *hipster* explanation of why he chose those tunes. When she read the list along with Chad's commentary, it brought her back to his carefree days and beatnik writing style. Chad influenced her musically. They liked many of the same artists. According to Teri, he was a music hound and avid guitar player. Wherever he went, to work or to be with friends in the big city, he always had his music and ear buds as constant companions. In college he was in a band but it wasn't really his deal to perform. He would just love to jam out with friends or play alone. From a young age, it was music that brought him joy.

His playlist became the audio mantra. St. Vincent, *Digital Witness* and James Blake, *Retrograde*, Pretty Lights, *Yellow Bird* along with Beck's *Blackbird Chain* among other tunes became a daily ritual. They were convinced he was hearing his music and was soothed by it despite his lack of consciousness. When using the patient's preferred music, this *does* help to comfort them.

Attempts at clamping the external ventricular drain (EVD) caused the intracranial pressures (ICP) to dangerously increase. The CAT scan was showing signs of hydrocephalus, the buildup of cerebrospinal fluid that can cause dangerous pressure on the brain. He was still heavily sedated. He required intravenous blood pressure medication. His left side wasn't moving and his left arm was very swollen. He began to shiver uncontrollably so aggressive attempts to lower his temperature ensued. His heart rate sometimes went too low or too high and irregular. Teri didn't remember sleeping much during this time. Averaging about 20 hours per day in the ICU, she volunteered to be an extra set of hands to help care for her son.

Each day that passed meant he would probably live but in what neurological state? He graduated Summa Cum Laude at American University in Washington, DC not even two years ago. Would that brilliant mind be lost?

The day before his 24th birthday, 12 days after his fall, Chad became more alert with his eyes fully opened for the first time! She felt the gentle squeeze of his hand. His eyes were beautiful, but they were confused and searching. They sensed he was very frightened. He couldn't speak so they spoke gently to him and explained in simple terms what had transpired. Teri promised they would not leave his side.

He turned 24-years-old in the Surgical ICU. A few days prior, they met the anesthesia resident. She and Chad had history because she was the resident on call the night of the accident and was a part of the surgical team. She brought in a helium balloon for Chad on his birthday. Over the course of the next two weeks she visited often, bought Chad milkshakes and cared for him exquisitely.

Teri and her daughter, Danielle were constant companions in the ICU with Chad. Danielle stayed overnight but Teri was fearful this was all too much. When Teri decided it was best for Danielle to stay back she protested, needing desperately to be with her brother. His other sister, Hannah was at the hospital constantly as well but the ICU was still toxic for her. She and her close friends back home were coordinating the care of the family home and dog in Cleveland. Hannah also helped read to him, played music, held his hand and spoke gently to him.

Both sisters cried and grieved over having to leave Chad and return home to their colleges but too much school was already missed. Teri cried openly when her family left NYC to return to their lives. Her only solace was being with Chad and clinging to the small signs of his re-emergence.

Visitors were a welcome relief. They came from Brooklyn and Manhattan. They drove, trained or flew from DC, up state NY, Cleveland and even as far away as California. Teri's friend, Cheryl, surprised her with a visit from

Rhode Island. Teri told me; *"each visitor was like a gold-plated package we unwrapped, savoring every moment of time with them."* A poster was made and still today, hangs in Chad's room at home as a reminder of their cheerful offerings.

The slow process of waking up from a coma is not done in minutes but rather over days and even weeks. Using Chad's playlist on Spotify, Bob Dylan sang to him constantly from albums Blonde On Blonde and Highway 61 Revisited. Much to their delight and surprise, with his new birthday socks on and barely conscious, Chad was tapping his right foot to the beat of Dylan. *This was truly one of the first signs of his revival. Not yet awake, he was responding to music!*

Prayers were heard. Tears of gratitude and joy were melting despair. They had no way of knowing where Chad was neurologically but each sign of life brought hope. From tracking with his beautiful eyes, to squeezing their hands, to one blink means yes, two blinks means no, every sign of life was seized.

As stated earlier, this is a slow process. The details of the brain re-awakening are complex and varied. As each day went by there was another sign of hope; the gradual weaning from the ventilator, recognizing faces of family and friends, nodding yes and no, writing short messages and hearing his voice speaking aloud. There were also times of great torment and terror. The scared look in his eyes, the inability to comprehend when or why he was in the ICU and why they were by his side, the disbelief when he heard them describe in simple terms the events of the past weeks.

Hundreds of conversations coordinated by health care teams from both NYC and Cleveland yielded a plan for transfer back to Cleveland to begin the long, arduous process of in-patient rehabilitation. Teri and Larry both expressed incredible gratitude for this coming together of both health care giants and the many involved in coordinating this transfer.

The next phase of Chad's recovery was about to begin. Saying goodbye to the familiar and adopting a new health care family back home was just a plane ride away! It started with an ambulance ride from Kings County to JFK Airport. Larry arranged for him to be transferred onto a private jet. It was captained by a calm, kind pilot, a friendly, competent RN and a paramedic. It was a clear, blue-sky day. They lifted off as Teri softly cried, sitting next to her miracle child, awake and talking, in awe of the last 28 days following the phone call that changed everything.

Teri remembered one of her favorite Bible verses that helped her get through this trying time: *"For I know the plans I have for you," says the Lord. "They are plans for good and not for disaster, to give you a future and a hope." Jeremiah 29:11* (New Living Translation Bible.)

Chad arrived at MetroHealth Rehabilitation Institute of Ohio on April 15, 2016..........

Since the other stories in this book took place, music therapy services were expanded to all the ICUs at the main campus of MetroHealth, two days per week. In order to do this, music therapy is now provided three days per week in the Rehabilitation Department rather than five as it had been in the previous chapters.

Shortly after they arrived, Chad's mother happened to be reading informational materials about MetroHealth's Rehabilitation Center. She caught notice of the fact that music therapy services were available. She was very excited and asked Chad's physical therapist (PT) about how he could receive it. The next day, Chad's PT was in a team meeting with the music therapy intern and mentioned to her Chad's desire to participate. This intern went to meet Chad and his mother to discuss the benefits of music therapy and explained our schedule. She clarified that we would ask his doctor for a referral. When my intern informed me that we would be receiving another musician as a new referral, I read his medical history to understand the details of his injuries.

On April 20, Chad was brought to his music therapy evaluation in a wheelchair by his mother Teri. I was first struck by how good he looked after sustaining a fall from four stories high. *He fell on his head, on cement!* What a miracle he was alive and so able bodied. Chad was wearing a helmet to protect his brain since he had a craniectomy but was not ready for the piece of skull to be replaced. Prior to being admitted to rehab the trach was removed. He responded to our greeting in a soft-spoken voice, very pleasant and cooperative. His mother was right by his side and happy for him to begin his journey in music therapy knowing how much music was a part of his life.

Chad and his mother explained his history in music to me. Chad had been playing guitar since he was eleven years old. He had one year of formal lessons but then continued to play on his own and in the church youth group. Prior to guitar, Chad played the viola in school for five years starting at the age of nine. He learned to read music for the viola. Obviously his viola and guitar playing overlapped and eventually the guitar won out as his desire to play viola diminished. Although he started to learn how to read music for the guitar during his formal lessons, this soon changed into reading tablature and chords. He played in a band during college for three years called "Dealers Hand", a Rock band with a Blues influence. They played in bars and at American University in DC, Chad's alma mater. Chad enjoyed covering musicians such as Ryan Adams, Bob Dylan and Neil Young. Chad had played guitar for 13 years by now, however, since his accident, he was not able to use his left hand. It had been five weeks since he could not functionally use his left hand or play guitar. Chad was desperate to play again.

During our orientation questions, Chad was able to state his full name, his birthdate, the current date, the name of the hospital/rehab center and fully understood what had happened to him. His mother explained in more detail that he was standing on a fire escape platform with his friend Joey and leaned against a railing which gave way. I learned later that most apartment dwellers in Brooklyn and NY City use their fire escapes as quasi balconies.

This all continued to be amazing to me. He spoke of the genres of music he enjoyed; Rock, Jazz, Folk, a little Country and specific artists he liked listening to such as Miles Davis, Pat Metheny, Bob Dylan and Neil Young.

I gave Chad our acoustic guitar and asked him to tune it for us. Again, this was one of my tests to recall his early training. Although he could not use his left hand, he plucked the strings and turned the pegs all with his right hand very carefully. Chad knew how to tune his instrument. This process took a while. He could name all six strings in order, correctly; "E, A, D, G, B, E" pointing to each one as he identified them. He followed all of my instructions, although slowly. His long-term memory was there but he had some short-term deficits. He had difficulty recalling new things he learned in the session, by the end of the session, when tested.

Since Chad was in the ICU for so long and his voice was so soft, I felt it was important to test his respiration or pulmonary strength. We asked him to sing or hum one note/syllable for as long as he could in one breath. He could sing "ma" for 11 seconds. This wasn't terrible but he needed work in this area to help prevent pneumonia and provide increased pulmonary strength which would also help him to feel better and increase his endurance.

Chad also suffered from headaches and during our session, his head was hurting. He rated the pain to be a seven out of ten on a self-reporting scale with one being no pain and ten being unbearable.

When his evaluation was completed, together we came up with a treatment plan. Obviously his first desire was to work on his left-hand motor skills in order to play his guitar again. I encouraged singing his favorite songs to increase his respiration and by being engaged in his favorite music, hopefully his pain would subside. Research has shown that when patients are involved with their preferred music, the brain is busy processing the music and therefore is distracted from the pain so they perceive less discomfort.

Our favorite music can also increase those "feel good" hormones as discussed before and natural brain chemicals that reduce pain.

Chad and his mother were in agreement with the music therapy plan and were excited to continue.

As Chad was leaving our first session he stated "I'm really excited about playing." I explained to them how the many musicians with whom I've worked through the years recovered well. I informed him that he would have to work very hard in music therapy to regain his guitar skills. He said he was willing to do the work. Chad and his mother were very thankful and happy to begin. Of course, no one really knows how much work it will take until they begin their process. Every injury is different; every brain recovers differently. Given my experience in rehab and in working with musicians after trauma, something inside told me he would recover. He demonstrated potential but I just wasn't sure how long it would take.

I still could not believe my eyes and ears. This was a young man who was not only still alive after a terrible, life-threatening fall, but was having appropriate conversations, had great long-term memory skills, looked great even with a helmet on and in a wheelchair, willing and motivated to work. What a moment for me. He and his family were very blessed.

During his five sessions in music therapy, until May 2, 2016, he progressed as demonstrated by the following:

1. Chad could play chords on the guitar but very, very slowly and needed moderate assistance from me to place his fingers in the appropriate positions. I had to hold his position on the guitar. This was strictly a motor planning issue and not a memory problem. He remembered how to finger all his chords by explaining where each finger should be placed.

2. His respiration improved to 14 seconds from a baseline of 11. This was very good considering most of our time was spent on left-hand motor

skills. He and his mother loved to sing the same music together. We had fun!

3. Chad's pain decreased in 50 percent of his sessions, down to a level four out of 10, much more manageable than his first report of seven out of 10. Chad's skull had not been replaced yet and headaches still arose from time to time.

We used the *Therapeutic Instrumental Music Performance* technique to improve strength and fine motor skills as well as motor planning ability in his left hand. Using his favorite music, I worked Chad's left arm and hand through exercises being guided by the rhythm. After I exercised his arm and fingers I asked Chad to try and move them without assistance from me. We saw some movement in his arm but very little with his fingers. I needed to provide most of the assistance. We used a variety of small percussion instruments in many different ways, to a variety of Chad's favorite genres of music. However, again I needed to hold his left hand around the instrument so he wouldn't drop it. We were desperately trying to get Chad's neurons which controlled his left upper extremity to fire and connect again.

When he was ready, Chad also worked on strumming and pressing the buttons on an autoharp with his left hand. Although slow and arduous, this was a strengthening exercise and helped to prepare him for the difficult task ahead, playing the guitar. The left-hand chord positions were especially intricate and required extremely fine motor skills. The delay in his left upper extremity recovery was most frustrating for Chad. Apart from his short-term memory, every other skill came back quickly. This made things especially difficult for him because he couldn't understand why his left-hand motor skills were taking so long to return! Nonetheless, he kept working and struggling because he knew it would all be worth it.

One positive intervention for Chad and Teri was singing together. They both liked much of the same music and artists so when we all sang to help

Chad improve his respiration, this became a nice family event for them while in music therapy. I remember singing *Mr. Tambourine Man* by Bob Dylan. We laughed because each verse seemed a little longer than the one before. Chad also chose songs such as *Old Man* and *Heart of Gold* by Neil Young to which we sang and exercised.

We had a nice time discussing his favorite artists and songs and music therapy became an enjoyable experience for both of them, as well as a place for him to start to regain his music skills. During these precious moments his perception of pain would decrease or he would totally forget about it at least during the music. His pain would most likely be eliminated when his cranioplasty was complete.

Teri asked if she could bring Chad's guitar in for him to practice with during anytime outside of the music therapy session. I said *"sure, I actually encourage musicians to practice on their own if they are able."* Unfortunately, their guitar time in Chad's room did not work out as well as they had hoped. Chad became easily frustrated with his lack of ability and his mother was at a loss as to how to keep him going and motivated. His motor skills were still lacking. When Teri explained this to me I told her not to force it and only try it for a few moments at a time. She was more successful with getting him to play egg shakers or claves but only for a few moments at a time.

One day my intern and I were in the elevator with two other women. I guess because of our conversation, they asked us if we were from music therapy. We confirmed we were. They explained that they were Chad's aunts and heard all about music therapy while visiting with him and they were so excited that Chad was involved with us. Then they offered us left over, homemade cookies they made for Chad and his family. OH MY WORD, those cookies were great! What an elevator ride that was. Chad didn't tell me about the other talents in the family!

When I went to get Chad from his room for his sixth and last session, his rehab physician (physiatrist), Dr. James Begley was there and said,

"I cannot keep him here any longer, he's doing quite well." Chad's mother looked at me and said, *"we are being discharged now."* Because Chad was walking so well and getting around by himself, his insurance required that he be discharged and receive outpatient therapy once or twice per week. He was discharged on May 2, 2016.

Just prior to discharge Chad's mother told me she was very interested in retaining my services privately. They wanted and needed more music therapy along with their other outpatient, physical, occupational and speech (cognitive) therapies. When she found out that we lived in the same city and that I would come to her home to treat Chad, she was very relieved and happy. We only lived 12 minutes apart! What are the odds?

On May 6, Chad had his first private, music therapy session. Chad and his mother both greeted me at the door in their normal, friendly manner. They had a beautiful home with a piano in their lower level along with bongos, egg shakers, claves and of course an acoustic and electric guitar. They were ready for music! Even their little fourteen-year-old, Pomeranian-Poodle Lilly, wanted to be part of the event.

Two Fuzzy Faces

Originally I suggested he receive music therapy two times per week but I understand that this can be costly for people since it is not usually covered by insurance. Given all they went through, I allowed them to make their own decisions as to whether they received one or two therapy sessions per week. They chose one. This was a change from receiving three per week down to one and receiving other therapies five times per week down to one as well. That's the way their schedules worked out with the other therapists.

I had Chad work on basic strengthening and fine motor skills using rhythm/music and small instruments as well as playing with one finger, one note at a time on the piano using his left hand. Depressing the keys on a real piano versus a keyboard, helps to strengthen the smaller muscles. At first we noticed only minor improvements but we took whatever we could.

One week in May, as I was leaving their home, Chad's mom gave me a box of those famous, homemade cookies from their relatives. My eyes must have opened wide and I said, *"Oh, I get a bonus today?"* We all laughed. I thought about sharing them with my college age son who was home for the summer but I ended up eating all of them on my way home. I didn't know what was in them and I didn't care, they were so addicting!

At the beginning of May, during our first few sessions in his home, Chad started to improve even more. Each finger, although slowly, was able to play the keys on the piano, one at a time. Although I had the metronome set for only 47 beats per minute, he could not keep up but was using a few fingers on his left hand nonetheless. He could play some hand-held percussion instruments and worked on stretching exercises for his left hand and full body. We also worked on proper posture when playing instruments or when at the piano. We attempted the guitar but I needed to place his fingers into position. His brain was still not connecting.

During the last two or three weeks in May, just prior to Chad's cranioplasty, we all noticed that he was regressing instead of progressing. The fingers on his left hand were more contracted, expressing too much tone, making it difficult for him to straighten his fingers or stretch them out. They became claw-like which interfered in his playing ability even using just one finger at a time on the piano. He could no longer isolate each finger. His whole hand went down at one time. His apraxia, the inability to plan purposeful movement due to brain damage, was becoming pronounced. The messages from his brain to his hand were disrupted. His intention tremors, which only occur during purposeful movement, were more pronounced. His mother was afraid it was because of a lack of therapy or the abrupt decrease in therapy time, even though she was very good about making sure he followed through with all of his daily homework assignments.

Alert and awake after his June 2 surgery at MetroHealth

Finally, Chad was scheduled for his cranioplasty, June 2! We all hoped that after this needed surgery, he would show improvement and would feel better too. He was still suffering from occasional headaches. Chad remained in the hospital after surgery for approximately one week prior to returning to the rehabilitation center where he resumed all his therapies. Everyone wondered how he would respond to yet another surgery. Nothing is guaranteed. All surgeries are a risk. Sometimes there are set backs and sometimes there are none. Chad experienced some setbacks which is why he was re-admitted to the rehabilitation center. They were hoping he could have been discharged directly home but they were happy he would receive therapies five days a week again.

On Friday, June 10, I evaluated him in music therapy at MetroHealth Rehabilitation Institute of Ohio. We knew his history and given the fact he was expected to remain in rehab for only 1½ weeks, we just focused on his left-hand skills. It was difficult and frustrating for him so I needed to explain that this sometimes happens after surgery. Surgery may help move a person forward but it also may temporarily move a person backward. His brain had been through hard times.

Chad still could not hold or play small, rhythm instruments with his left hand. His apraxia continued and he still demonstrated too much tone along with intention tremors. We had to start all over again which was disappointing for everyone but I needed to remain positive and strong for him and his family.

Now he would need to have patience like he never experienced before. This was going to be a test of willpower, discipline and his desire to play guitar again. I must say he has been one of my most challenging musicians. On the other hand, although many of them experienced life threatening situations, none of them fell 40 feet from a building onto cement! Honestly, I don't know if I could have done what I was about to put Chad through the next several weeks. This was not going to be easy. I knew I had to be able to motivate him, keep his spirits high, be honest with him about his progression each week but discuss every little spark of improvement and point them out to him. Chad was very hard on himself, criticizing every little thing he could NOT do. He had to learn NOT to compare himself to how he played guitar before his injury but compare himself to how he played in his last session one week ago. He really needed to learn the "one day at a time" philosophy and celebrate every little, positive step of improvement.

Given the fact that music therapy was offered three times per week in rehab, our goal for the first week was just to be able to grasp and have enough strength in the left hand and arm to play some small rhythm instruments like claves, egg shakers and maracas with moderate assistance from me. That meant in order for him to play those instruments he would need me to do 50 percent of the work holding his hand and arm.

Chad only participated in four music therapy sessions after his evaluation on June 10. After two sessions, using all the techniques I could possibly use with him, he was only able to do "the one finger pinch" where his index finger and thumb could pinch together. With assistance from me, he could hold but not play some rhythm instruments.

Chad learning to play claves using an Ace bandage to build up grip

Supporting his left arm on his lap he could lift his wrist from lap level but not the rest of his arm. He was not able to do any of these things when we first started. However, in order to actually play the rhythm instruments, I needed to fully move his left arm for him. We did all these things to his favorite music and sometimes sang along to make it more fun and interesting.

Teri tried to encourage Chad to continue these music exercises in his room either in the evenings or on weekends and during the days when music therapy was not available. Sometimes he was in the mood and sometimes not. Chad was anxious for his arm to work as it did prior to his injury.

After another two sessions Chad could play two small instruments independently with no help from me! We built up the claves with an Ace bandage wrapped around it to make the grasp bigger and easier to hold. He could play some of the other instruments with moderate assistance from me but at least, not as much as before. These instruments were heavier to hold and had a larger grasp. We had to notice every little spark of progress to demonstrate to him that he was moving forward.

As you might understand, this took a great deal of work and patience. At that point in time, Chad was walking around the unit, always spoke appropriately but had short term memory problems. He liked talking about some of his favorite musicians and the songs they wrote. His speech therapist gave him homework to turn in for when they were going to meet each other on the outpatient side after discharge. Chad had to make a list of 20 protest songs for his homework. He enjoyed doing that and it was a nice assignment he and his mother, a former guitar player herself, could do together.

After his cranioplasty and prior to his second discharge home on June 21, Chad participated in four music therapy sessions and one drum circle. Finally, his family could take him home for good! Although they missed having therapy every day, they were excited to be home all together as a family again. It had been more than three months since his accident.

On June 29, eight days after discharge, I was back at the Millers' home providing private music therapy sessions to Chad. We sometimes began our sessions with very upbeat music to which I had Chad dance and move to the rhythm so his entire body would be primed and sync with the beat. This was purposeful because it basically says to the brain *"wake up."*

He continued to play the built-up claves like the ones we used in the hospital. Sometimes he could play the egg shakers independently but they were slippery and would fall out of his hands. I also brought him a homemade frame drum; a square wooden frame wrapped tightly with packaging tape. He could hold a drum stick and play the frame drum, alternating his right and left hands, only to give his left hand a rest at times. His rhythm was great, he almost looked like a drummer. Of course his left hand was weaker than his right but it was improving in coordination, grasp and strength. As a result, we started to incorporate his electric guitar. The strings on the electric were easier to depress than those on the acoustic. Chad would still use his right hand to place his left hand fingers in the position of each chord. Even holding the guitar in its proper place was difficult because his left arm, although getting stronger, was still too weak to hold up the heavy, electric guitar. Because of this, I would have Chad sit back on the big, comfy couch they had in the lower level and prop up his left arm with blankets or pillows. This worked for a while but then I had another idea!

We went back to using the acoustic guitar but would lie it flat as if it was a lap guitar. I had Chad use one finger on his left hand, press one string and then pluck that string with his right hand which would tell me if he was pushing down hard enough with his left finger, showing increased strength. He would slide that left finger down the neck of the guitar and at each fret would pluck again. I had him repeat this procedure with each finger on his left hand, using different strings. This was isolating each finger and it alleviated the necessity for lifting and playing the guitar at the same time, all with the left arm and hand. Chad liked this idea. He could do it! It was like playing a table top guitar for those of you readers who might remember them.

Chad would practice this procedure and also played his small percussion instruments to music. His left hand was becoming stronger although he had difficult times once in a while. Chad started with shorter practices during the week; 10 to 15 minutes. He was attending all his other therapies as well, usually once per week. From time to time, Chad would play the acoustic guitar in its normal position but would still use his right hand to place the left hand where he wanted the fingers to be. I allowed this for a while but when I didn't see improvement in his left fingers, I told Chad that he shouldn't do that anymore. By using his right hand to place the left hand where it should be, he wasn't allowing his brain to do the work. We needed to make the brain tell the left hand what to do. So a big change in procedure was required, a very difficult one.

I told Chad that he had to "tell" the fingers on his left hand where to go on the guitar and wait for it to happen. We were going to take one finger, one string, one fret at a time and pluck that note when his hand was in place. So Chad actually verbalized, *"the index finger will play the D string at the first fret."*

Then the hard part was waiting for the left index finger to position itself without using the right hand to put it in place. He had to retrain his brain. It was slow but it worked. I'm sure it was excruciating for him. He did this with all the fingers on his left hand. Some were better than others. I didn't measure the time it took to do this but it felt like it took two minutes for each finger to go to its directed place. I was amazed at how patient he was. His mother even said she didn't know if she would have the patience for this if she were in his place. It was difficult and tortuous but needed to be done. When they went to their next outpatient, occupational therapy appointment and explained what I required of Chad, their therapist said, *"yes, that's exactly what he has to do!"* She supported my recommendation.

Chad continued to practice this method and over the weeks we began to see hope. Each finger was doing its job, playing one string, one fret at a time. At that point, his fingers moved so slowly that we were unable to incorporate the metronome but eventually we did. The metronome was started at 34 beats per minute but Chad needed to take four counts to place each finger where he wanted it to go to pluck one note. Then he started to play two fingers at a time, again counting four beats before placing them into position. He couldn't accomplish too many of the two finger positions but this was an improvement.

I asked Chad to document his practice sessions in a notebook because when I arrived for his private sessions and asked how things went during the week, he would often say, *"not so good"* or *"it was ok"* because he wouldn't see the improvements he made. Chad didn't remember the small but detailed improvements that occurred in the week. After the notebook method began and he kept track of what he did, I could point out all the new things he was doing. He could now play many two note combinations and play eight easy chords; CM7, A7, A, E7, D, D7, E minor and C. We also increased the tempo to 46 beats per minute and instead of placing a finger on a string every four beats, he eventually decreased it to two beats. Again, this is a way to measure his speed and dexterity in his left hand.

At this writing, Chad can play all the percussion instruments in his house with no build-up for grasping and does not drop any of them. His grasp and fingers are stronger and his posture is better, holding the acoustic guitar properly without assistance! It's still difficult for Chad who is quite often down on himself. According to his mother, this is his nature. They started rating Chad's practices because of the conflict in their perception of his progress. One week, Chad rated his practice sessions as being a five out of 10 but his mother rated the same week as an eight out of 10. He is able to do new things but it's still not enough for Chad. He practices a full 30 minutes or more per day doing chromatic exercises, two note combinations and beginning chords using a metronome. Just remember, in June he could not hold any small, hand-held instrument without my help. The guitar is one of the most difficult instruments to use for small muscle control but a good one! It's the instrument he wants to play again.

For the last few months Chad has also been jogging two miles per day with one or the other parent. He works out at the local recreation center, takes his turn once per month controlling the audio system for his church worship bands on Sundays and is now taking a creative writing course at the local community college. My how he has progressed! Chad's hair is also growing back and he's looking more and more like the pictures I see around his house.

Chad, you're still the good-looking guy I met in a helmet! Keep pluckin' away and the music will find you!

Chad practices his "two finger position"

"I started playing guitar when I was eleven-years-old and it pretty quickly became my greatest passion in life. Music therapy then, was an immediate draw for me; regaining the ability to play guitar was and remains incredibly important to me. It's been frustratingly slow moving but I can't thank Carol enough for all her help and insight."

Chad Miller

CHAPTER VI

Literature Review
The Comparison of Brains Between
Musicians and Non-musicians

This review is by no means comprehensive nor all inclusive. This area of study is voluminous with multiple areas of science publishing frequently; neurology, neuroscience, neuropsychology, cognitive science, and more. A review of the literature, which explains how musicians' brains are different than non-musicians', provides possible explanations as to why I believe musicians might recover more quickly and more efficiently than non-musicians, post trauma especially **when using music therapy in their rehabilitation**. Many researchers believe that extensive, music training can be a "protectant" for the brain.

Numerous studies over the last decade or longer discuss how music training is a multi-sensory experience that recruits and challenges many regions of the brain simultaneously (Gooding et al., 2013, Barrett et al., 2013, Seibert et al., 1999). This is what makes training in music so unique to other forms of cognitive training as well as other art forms. It requires several neural-networks to synchronize and work together. Much like physical exercise is a workout for the body, we now know that musical training is a workout for the brain (Collins 2014). Playing an instrument or learning to sing requires abilities such as; good auditory perceptual skills, planning and executive functioning, fine and/or gross motor activities, emotional expression, reading musical notation (a symbolic system) translating that system into a motor response and memorization (Schlaug et al., 2015)

We know that music training promotes neuroplasticity. Dawson (2011), provides an extensive although not comprehensive bibliographic review of articles depicting neuroanatomic and functional differences between musicians and non-musicians. There are structural changes that take place in the brains of musicians vs non-musicians, especially in the auditory and motor areas and these changes can last well into a person's older life (Wan and Schlaug, 2010, Kraus & Chandrasekaran, 2010). Kraus & Chandrasekaran (2010) found these structural changes beginning to take effect even with only 15 months of training.

White-Schwoch et al., (2013) found that a moderate amount of music training, i.e.; four to 14 years early in life is associated with a quicker response time in relation to speech later in life even long after training stopped. These authors suggest that early music training changes the connections in the brain that respond and interact to sound. Over time these interactions may create sharpened neural processing in the auditory system well into older age. Regarding auditory processing and speech and language abilities, Kraus & Chandrasekaran (2010) found that musicians have enhanced skills in these areas as well as cognitive and sensory abilities. They explain that musicians have a stronger auditory system with increased verbal memory and auditory attention abilities especially in challenging listening environments.

Wan & Schlaug (2010) discuss and display brain imaging that depicts a larger arcuate fasciculus in the brains of musicians as compared to non-musicians. This is the neuro-fiber connection between the auditory and motor systems where sounds are perceived and then followed by a motor response. This same neuro-pathway is the connection between Wernicke's and Broca's areas in the brain. These are the two speech centers where words are understood in a meaningful way and then speech is produced respectively. This might explain why in my practice, many musicians have milder speech production deficits if any, after brain injury.

One study examined the differences in physical, sensory and motor parameters in Writer's Cramp versus Musician's Cramp. Due to overuse, the musicians in this study suffered dystonia, another way of saying Musician's Cramp. After a period of rest and then a particularly designed re-training program, musicians made a better overall recovery in upper extremity skills compared to the non-musicians (McKenzie, et al. 2009).

Wan & Schlaug (2010) and Collins (2014) describe the changes in the corpus callosum that take place after music training and how this area is larger in the brains of musicians. This is the bridge between the two hemispheres of the brain that carries electrical impulses back and forth facilitating communication between both sides of the brain. In musicians, this can take place faster and through more routes because of the enhanced neuro-connections.

Galarza et al., (2014) reported a case study of a virtuoso jazz guitarist with epilepsy and severe behavioral disturbances. This musician underwent a lobectomy where 70 percent of his left temporal lobe was removed resulting in severe retrograde amnesia and the loss of all his music skills along with many other abilities. He was gradually introduced to his own recordings of music which helped his biographical memory to return in two years. His father once again taught him guitar, as he did the first time and in several years, he regained his music skills even at the virtuoso level. The amazing thing in this case study report is the fact that the temporal lobe is one of the four major lobes of the cerebral cortex and is responsible for processing memory, emotions, hearing, language and learning. These skills are necessary for good music making. The left temporal lobe is where the primary auditory cortex is housed which processes both speech and vision. In addition, Wernicke's area spanning the temporal and parietal lobe, is an area in the brain that is key for speech comprehension.

Seibert et al., (2000) describes a case study of a 20 year-old college musician who was an aspiring professional oboist. While wading across a river she lost her balance and was swept away suffering profound injuries including hypothermia, cardiac arrest and severe hypoxic brain injury. This article details her injuries and fight back to recovery using music in the process and includes quantitative and qualitative data to analyze the powerful role of music.

This young, college musician like the jazz guitarist mentioned above, did not have a qualified music therapist providing treatment after their injuries but had a variety of music and musical activities provided for them by relatives and friends. I question whether they could have recovered sooner had they received music therapy services from a Board-Certified Music Therapist.

Regarding cognition, Gooding et al., (2013) provide evidence that music training helps to improve cognition later in life even well after musical training has stopped. They found the ability to read music was associated with different types of memory and verbal fluency. They propose that early to midlife musical training may be associated with improved episodic and semantic memory later in life as well as a useful marker of cognitive reserve; brain structure differences that may increase tolerance to pathology. Hanna-Pladdy & MacKay (2011) provide a correlational study that suggests cognitive functioning in advanced age might be preserved with active music making throughout one's life time. When musicians actively played music for 10 years or more, they demonstrated higher cognitive functioning much later in life even years after they stopped playing. Length of time in active music making correlated with better cognitive reserve later in life. Sluming et al., (2002) found that orchestral musicians do not demonstrate age-related volume reductions in the brain like their non-musician counterparts. Their study only included males.

Jacobsen et al., (2015) discovered an area in the brain that has been identified to store long term musical memory. Describing the three bio-markers of Alzheimer's disease, this article explains how the long term musical memory area of the brain is not affected by these bio-markers at least until the final stages of the disease. It is remarkable to observe how people affected with severe Alzheimer's disease cannot remember much about themselves, their families or other details but they often can remember the lyrics and melodies to the old tunes they grew up listening to. This article did not discuss musicians but the benefits of long term music memories.

Schlaug (2015) provides a basis, using the musicians' brain as a model for why music can and should be used for neuro-recovery in the area of brain injury or developmental disorders. He goes on to say that plasticity as a result of active music training suggests the potential for music interventions in neuro-rehabilitation particularly in the area of vocal-motor functions.

Omigie & Samson, (2014) suggest three hypothesis how musical training prior to brain injury can bring about preserved musical and language abilities post injury; a) anatomical differences in the brain of a musician b) greater redundancy allowing access to different strategies and c) greater metaplasticity in the brain of a musician. They compared the results of case studies from the year 1900 and on of professional musicians that either had cerebral vascular accidents, tumor resections or epileptic resections. Twenty out of 35 musicians lost their musical abilities post injury. Half of that number experienced language deficits and half did not. The musicians who experienced a better overall recovery were those that had the resection of tumor or for epileptic treatment. None of their examples were traumatic brain injuries like the stories in this book. None of their examples reported receiving music therapy post injury or surgery. When looking more closely at the musicians in this study since the year 2000, 11 out of 15 did well, meaning they recovered in at least two out of three parameters measured; receptive music skills, expressive music skills as well as language skills. Comparing musicians from years past is difficult due to lack of updated therapy techniques and medical advances of today.

Due to modern imaging technology and the vast interest in music's extensive effect on the brain, scientists are beginning to ask, "can music training protect the brain from disease?" Others ask, "can music training aid in a more efficient or qualitative recovery process after trauma?"

Conducting randomized clinical trials comparing musicians' and non-musicians' recovery process after brain injury is needed to support my hypotheses.

CHAPTER VII

Similarities in Recovery of Musicians with Whom I've Worked

Over the last 12 ½ years working primarily in the Rehabilitation Department, I have worked with numerous musicians from the high school level on up to professional; both instrumentalists and singers. Using evidenced based music therapy techniques to help them recover from compromised motor, cognitive or speech and language deficits helped most of them to return to playing their instruments or singing even prior to leaving the hospital. Many returned to school or work as well.

This book only highlights the recovery process of a few musicians with whom I have maintained contact and who have given back to MetroHealth by performing at our special events since their injuries. In working with musicians, I see some similarities in their recovery process and I would like to highlight them here. This is by no means based on a research analysis of data but from my own professional experience, opinions and records. My hope is, this will help with future research endeavors.

Speech and Language: Many musicians with whom I have worked seem to have very mild expressive, speech deficits or none at all after brain injury. The speech issues I mentioned in these case studies were so mild I did not have to address them or addressed them very little. Could this be because of the thicker, fuller arcuate fasciculus mentioned in the literature review which precedes this section? As I explain in the literature review, the arcuate fasciculus is the bundle of white fibers that connects the temporal, parietal and frontal lobes or the auditory system to the motor system. It aids in messages sent between the speech centers as well. I was actually amazed at the ability of these musicians, no matter the type of brain injury, to express their needs and communicate with others.

Lower Extremity Motor Skills: Although musicians need to have physical therapy as well as all other therapies as indicated by their respective injuries, I find that musicians, comparing brain injury to brain injury, seem to recover lower extremity motor skills sooner than their non- musician counterparts. This may again be due to the stronger, more enhanced arcuate fasciculus in the brains of musicians. Walking is a temporally based movement simulating music in 2/4 or 4/4 meter. This happens naturally for all of us but especially for musicians. In addition, the anterior half of the corpus callosum in musicians' brains is larger allowing for more frequent messaging between hemispheres. Motor related structures in musicians' brains are enhanced as well. This might explain why they seem to recover lower extremity motor skills sooner.

Pain: When musicians can listen to or engage in their preferred music without physical effort to play (meaning without upper extremity motor deficits to play an instrument or vocal deficits to sing) they seem to perceive much less pain *or no pain at all.* Their brains seem to concentrate more fully in analyzing the music, therefore cannot neurologically process the pain. I have even found this to be true with those who strongly appreciate music although not musicians. They seem to benefit by the music to reduce pain more significantly than others. We also know that our preferred music increases the neurotransmitters that improve mood and activate the reward centers in the brain making us "feel good." There are many good pain studies that use music for treatment in the research literature.

Respiration: Avid singers and wind instrumentalists either have very good respiration skills or recover more quickly in this area due to all their pulmonary exercise through singing or playing a wind instrument.

REFERENCES

About Education: Temporal Lobes (2016) Retrieved from: http://biology. about.com/od/anatomy/p/temporal-lobes.htm.

Brain Injury Association of America, Brain Injury Statistics (2015) Retrieved from: http://www.biausa.org/glossary.htm.

Collins, Anita (2014) How Playing an Instrument Benefits Your Brain. Retrieved from: https://www.youtube.com/watch?v=R0JKCYZ8hng.

Dawson, William (2011) How and Why Musicians Are Different from Non Musicians – A Bibliographic Review, Medical Problems of Performing Artists; 65-78.

Galarza et al., (2014) Jazz, Guitar and Neurosurgery. World Neurosurgery 81 (3/4): 651.E1-651.E7. Retrieved from: http://dx.doi.org/10.1016/j. wneu.2013.09.042.

Gooding et al., (2013) Musical Training and Late Life Cognition. American Journal of Alzheimer's Disease and Other Dementias. 201X, Vol XX(X): 1-11.

Hanna-Pladdy & MacKay (2011) The Relation Between Instrumental Musical Activity and Cognitive Aging. Neuropsychology. Vol 25: No.3: 378-386.

Jacobsen et al., (2015) Why Musical Memory Can Be Preserved in Advanced Alzheimer's Disease. Brain. 1-13.

Kraus & Chandrasekaran (2010) Music Training for the Development of Auditory Skills. Nature Reviews Neuroscience 11, 599-605.

McKenzie et al., (2009) Differences in Physical Characteristics and Response to Rehabilitation for Patients with Hand Dystonia: Musicians'

Cramp Compared to Writers' Cramp. Journal of Hand Therapy. April-June, 172-181.

Omigie & Samson (2014) A Protective Effect of Musical Expertise on Cognitive Outcome Following Brain Damage? Neuropsychology Review – Springer 24: 445-460.

Schlaug (2015) Musicians and Music Making as a Model for the Study of Brain Plasticity. Progress in Brain Research. Vol 217: 37-55.

Seibert et al., (2000) Music and the Brain: The Impact of Music on an Oboist's Fight for Recovery. Brain Injury. Vol 14: No.3, 295-302.

Sluming et al., (2002) Voxel-based Morphometry Reveals Increased Gray Matter Density in Broca's Area in Male Symphony Orchestra Members. Neuroimage 17(3): 1613-1622.

Wan & Schlag (2010) Music Making as a Tool for Promoting Brain Plasticity Across the Life Span. Neuroscientist. (5): 566-77.

White-Schwoch et al., (2013) Older Adults Benefit from Music Training Early in Life: Biological Evidence for Long-Term Training-Driven Plasticity. The Journal of Neuroscience. 33 (45): 17667-17674.

APPENDIX I

Brain Injury Statistics In America

To set the stage of how prevalent brain injury occurs in the United States, according to the Brain Injury Association of America, an estimated 2.4 million children and adults in the U.S. sustain a traumatic brain injury (TBI), caused by an external force and another 795,000 individuals sustain an acquired brain injury (ABI) from non-traumatic causes each year. These statistics do not include birth defects.

Types of Acquired Brain Injury Estimated Annual Incidences:

- Stroke: 795,000
- Tumor: 64,530
- Aneurysm: 27,000
- Viral Encephalitis: 20,000
- Multiple Sclerosis: 10,400
- Anoxic/Hypoxic: No National Data Available.

Currently more than 5.3 million children and adults in the U.S. live with a lifelong disability as a result of TBI and an estimated 1.1 million have a disability due to stroke. (http://www.biausa.org/glossary.htm)

APPENDIX II

MetroHealth Trauma Center

The Level I Adult Trauma Center and Level II Pediatric Trauma Center at MetroHealth Medical Center are here—ready to help—when you need us most.

The MetroHealth Trauma Center is one of the busiest in the nation. Trauma really is our specialty.

MetroHealth was one of the first verified trauma centers in the area. We were verified in 1992.

Each year, (per 2015 statistics) we saw approximately 6,000 trauma patients. That's more than 16 trauma patients per day, every day.

The average, injured trauma patient will see between 50 and 100 specialists during the course of their care at MetroHealth.

The Centers for Disease Control and Prevention state that getting care at a Level I Trauma Center like ours can lower your risk of death from trauma by 25 percent.

There are medical emergencies and then there are traumatic, life-threatening injuries. These are the conditions that can remain with you long after the event is over. These are the wounds that might need lifesaving care in the moment and long-term care as you recover.

At MetroHealth, people of all ages are brought to our trauma center due to:

- Falls
- Car crashes
- Gunshot wounds

- Blunt trauma
- Stab wounds
- Brain injuries
- Burns
- Additional complex injuries

These life-threatening injuries often require more care: more surgeons, more physicians, more equipment, more nurses and more expertise than a standard emergency room can provide.

The trauma center also works with:

- Orthopedic surgeons
- Neurosurgeons
- Plastic surgeons
- Ear, nose and throat (otolaryngology) specialists
- Thoracic surgeons
- Interventional radiology specialists

The team stands at the ready to provide comprehensive care to anyone who has a traumatic injury.

Retrieved from www.MetroHealth.org

ABOUT THE AUTHOR

Carol Shively Mizes, MT-BC

Carol Shively Mizes is a Board Certified Music Therapist and currently works as the Coordinator of Arts Therapies in the Department of Arts in Medicine for The MetroHealth System in Cleveland, Ohio.

Carol has been a Music Therapist for almost 35 years starting in the area of Long-Term Care with more than 20 years of experience working with people who had Alzheimer's disease as well as other forms of dementia.

Carol developed and implemented the first Music Therapy program in several facilities in Northeast Ohio and also consulted for others.

In 2004 Carol designed the Music Therapy program for The MetroHealth Rehabilitation Institute of Ohio, treating patients who have suffered brain injuries, strokes, violent head trauma, and spinal cord injuries along with other musculoskeletal deficits. In 2015, Music and Art therapies were transferred from the auspices of the Rehabilitation Department to the Department of Arts in Medicine where both services can extend throughout the entire MetroHealth System. Carol has expanded music therapy services to the Medical, Trauma, Surgical, Cardiac, Step Down, Intensive Care Units and other medical floors while still treating patients in rehabilitation.

She has mentored many music therapy interns and assisted in designing a University Affiliated Internship Program first with Baldwin Wallace University and then with other universities who offer a major in Music Therapy.

Carol graduated Cum Laude from Cleveland State University in 1982 with a Bachelor of Music Degree and a major in Music Therapy from the Cleveland Music Therapy Consortium. Carol completed her advanced Fellowship training in NMT with the Robert F. Unkefer Academy for Neurologic Music Therapy.

As a sophomore at Cleveland State, Carol performed with the New York City Metropolitan Opera Company when they brought their production of "I Am The Way" to Cleveland. She also toured with Paul Anka as a member of the CSU Chorale in a group of back-up singers.

Carol has provided numerous presentations for a variety of professional organizations in North East Ohio regarding evidenced based music therapy techniques for recovery. She also co-authored an article regarding Alzheimer's disease and music therapy:

Shively, Carol & Henkin, Lauren (1986) Information Sharing - Music and Movement Therapy with Alzheimer's Victims. Music Therapy Perspectives. American Music Therapy Association, Inc. Vol. 3.